DAVID WILLIAMSON's first full-length play, *The Coming of Stork*, premiered at the La Mama Theatre, Carlton, in 1970 and later became the film *Stork*, directed by Tim Burstall.

The Removalists and *Don's Party* followed in 1971, then *Jugglers Three* (1972), *What If You Died Tomorrow?* (1973), *The Department* (1975), *A Handful of Friends* (1976), *The Club* (1977) and *Travelling North* (1979). In 1972 *The Removalists* won the Australian Writers' Guild AWGIE Award for best stage play and the best script in any medium and the British production saw Williamson nominated most promising playwright by the London *Evening Standard*.

The 1980s saw his success continue with *Celluloid Heroes* (1980), *The Perfectionist* (1982), *Sons of Cain* (1985), *Emerald City* (1987) and *Top Silk* (1989); whilst the 1990s produced *Siren* (1990), *Money and Friends* (1991), *Brilliant Lies* (1993), *Sanctuary* (1994), *Dead White Males* (1995), *Heretic* (1996), *Third World Blues* (an adaptation of *Jugglers Three*) and *After the Ball* (both in 1997), and *Corporate Vibes* and *Face to Face* (both in 1999). *The Great Man* (2000), *Up for Grabs*, *A Conversation*, *Charitable Intent* (all in 2001), *Soulmates* (2002), *Birthrights* (2003), *Amigos* and *Flatfoot* (both in 2004) have since followed.

Williamson is widely recognised as Australia's most successful playwright and over the last thirty years his plays have been performed throughout Australia and produced in Britain, United States, Canada and many European countries. A number of his stage works have been adapted for the screen, including *The Removalists, Don's Party, The Club, Travelling North, Emerald City, Sanctuary* and *Brilliant Lies*.

David Williamson has won the Australian Film Institute film script award for *Petersen* (1974), *Don's Party* (1976), *Gallipoli* (1981) and *Travelling North* (1987) and has won eleven Australian Writers' Guild AWGIE Awards. He lives on Queensland's Sunshine Coast with his writer wife, Kristin Williamson.

DAVID WILLIAMSON

INFLUENCE OPERATOR

TWO PLAYS

CURRENCY PRESS,
SYDNEY

CURRENCY PLAYS

Influence and *Operator* first published in 2005
by Currency Press Pty Ltd,
Gadigal Land, 46-56 Kippax Street, Surry Hills NSW 2010 Australia
enquiries@currency.com.au
www.currency.com.au

In accordance with the requirement of the Australian Media, Entertainment &
Arts Alliance, Currency Press has made every effort to identify, and gain
permission of, the artists who appear in the photographs which illustrate these
plays.

NATIONAL LIBRARY OF AUSTRALIA CIP DATA

    Williamson, David, 1942–.

        Influence; Operator: 2 plays.

        ISBN 0 86819 759 9.

        I. Williamson, David, 1942– Operator. II. Currency Press. III. Title. IV.
        Title: Operator. (Series: Currency plays).

    A822.3

Set by Dean Nottle
Cover design by Kate Florance
Front cover shows John Waters as Ziggi Blasko in the 2005 Sydney Theatre
Company production of *Influence* (photo: Heidrun Löhr); and Henri Szeps as
Douglas and Rory Williamson as Jake in the 2005 Ensemble Theatre production
of *Operator* (photo: Steve Lunam).

Contents

Currency Press acknowledges the Traditional Owners of the Country on which we live and work. We pay our respects to all Aboriginal and Torres Strait Islander Elders, past and present.

INFLUENCE

John Waters as Ziggi in the 2005 Sydney Theatre Company production.
(Photo: Heidrun Löhr)

David Williamson's *Influence* and Talkback Radio

Graeme Turner

When the Productivity Commission released the findings from its inquiry into the Australian broadcasting sector in 2000, it pointed out how little we know about the scale and the nature of the media's influence over the attitudes and opinions of their audiences. Assumptions about media influence are built into the regulations which structure our broadcasting industries; television is thought to be more influential than radio, for instance, with the print media in the middle. But at this stage no one really knows for sure.

Some people certainly *think* they know. In the late 1980s, radio was deregulated because it was thought no longer likely that any one station or broadcaster could exert significant influence on public opinion. Since that time, however, talkback radio has become overwhelmingly the medium of choice for politicians. When the Budget is handed down each year, the Treasurer and the Prime Minister hit the airwaves to promote it. Largely, these are radio airwaves. The print media tends to have to listen to radio to find out what the politicians are saying—as well as to what the public are saying in response. On television, coverage of the reaction to the Budget will largely be comprised of video footage of the Treasure or the Prime Minister in radio studios taking listeners' calls.

Why is this so? Politicians love talkback because they can deal directly with their electors; their comments will go out live without any editing; and they don't have to put up with the annoying questions they tend to get from journalists. When they have a choice, then, they use talk radio as the format through which to communicate with the public.

The Cash for Comment controversy also demonstrated what some of the nation's leading businesses thought about radio's potential to influence public opinion. There would be little point to the millions invested in some of the nation's leading radio personalities if there was no likelihood that their comment and opinions had an impact on their listeners' views on the world. While some in the media industries defended the Cash for Comment payments as a purely commercial transaction and therefore entirely uncontroversial, it did raise serious concerns about the power of these radio personalities and the level of responsibility with which this power was exercised. John Laws was correct to point out that he was hired as an entertainer, not a journalist, and so to ask him to observe the journalists' code of ethics in his on-air performance was unfair. On the other hand, many of the issues with which these personalities deal every day are ones where the community urgently requires accurate and responsible information. Those who are also journalists may well accept the responsibility to ensure they provide accurate information but, as Laws' comments explain, this is not necessarily the case for those who are not journalists.

Talkback has become a dominant format for commercial AM radio in Australia. It has effectively replaced radio current affairs programming for most of the commercial sector, with regular radio current affairs surviving only on one national commercial network and the ABC. The value of current affairs programming is that in many cases it is aimed at providing information responsibly; using sources in an ethical, impartial and unbiased manner; and observing the professional standards of journalism. That it has been displaced by talkback means that we have replaced information with opinion, journalism with entertainment, and—at its worst—responsibility with demagoguery. No matter how much authority the talkback hosts generate through their manner, their voice, or their personal self-confidence, none of that can compensate for the fact that their listeners are absorbing opinions which are no better informed than their own.

This is the situation that David Williamson's *Influence* addresses. His Ziggi Blasko is an entirely recognisable representation of a certain breed of Australian shock-jock. The issues Blasko deals with on his program are precisely those upon which the Australian community is most divided: immigration, 'political correctness', feminism, multiculturalism and national identity. Williamson's Blasko mines these divisions—not to resolve them but to exploit them as a genre of entertainment. That would not be too worrying if all we got out of it was a good laugh. However, Williamson uses the character of Zehra, the Turkish-Muslim housekeeper, to demonstrate that this is far from the actual effect.

At the core of *Influence* is the conviction that the broadcasting of opinion generates a social effect. Further, Williamson implies, the more aggressive and intolerant the opinion, the more destructive this effect is likely to be for the community. While it is important to acknowledge that most Australian talkback hosts would not seek to present such views, there are still plenty who do—and whose audiences expect them to do so. In this play, David Williamson suggests that it is precisely the most uncompromising opinion that is the most commercially desirable for the radio industry. Shades of grey are for the faint-hearted, and somehow it seems that truths are stronger if they brook no qualification of concession. In many of the contexts where public debate occurs in Australia today, positions are marked by the vehemence with which they are expressed rather than by the quality of the evidence they can mount in their support. Talk radio can be regarded as one of those contexts, where the host's power to control the debate all too easily modulates into simple, old-fashioned bullying.

A chilling aspect of Williamson's portrayal of Ziggi Blasko is the fact that Blasko seems genuinely to believe what he says on air. Far from the Hollywood model of the media figure who cynically chases his market by telling them what they want to hear, Williamson's Blasko is a bigot by conviction. He doesn't see what he does as merely entertainment: he sees it as his personal contribution to the formation of public opinion. Even though the events of the play expose him to the effects of his own opinions in ways that demonstrate their inadequacy, he seems immune to self-doubt. What sustains him throughout is the loyalty of his audience, the public who call in to support him. Confirmed in his positions by their sympathetic phone calls, he is able to claim that Australia will always want to hear what he has to say.

Unfortunately, some Australians certainly will. However, what Williamson's play highlights is the social and human cost of allowing such voices the power to put their views with so little in the way of community responsibility to constrain them. While there are many ways in which talk radio can build and create community, *Influence*'s Ziggi Blasko is an example of its destructive capacities: to fracture and divide a community. The media format doesn't do this by itself, however; it is we, ourselves, who have allowed it to be put to work in this way.

Graeme Turner is the director of the Centre for Critical and Cultural Studies at the University of Queensland. He is currently conducting a three-year study of talkback radio with funding from the Australian Research Council and media monitoring firm, Rehame.

Playwright's Note

David Williamson

We are a fearful species. Evolution seems to have wired us to be excessively afraid. John Howard won an election by elevating the flight of a few hundred desperate people to our shores into a major threat. A threat so grave, seemingly, that our armed forces had to board a ship of another country to control a sick, dehydrated and debilitated bunch of refugees going nowhere. Our emotional system is thus, presumably, because, on average, people who over-reacted to threats survived at a greater rate than people who under-reacted to them. The old dictum, 'Better to be safe than sorry', seems deeply embedded in our psychology.

The opinion makers who control our airwaves intuitively understand this fact about us. They understand how easy it is to generate fear by overstating and exaggerating supposed threats to our wellbeing and to turn that fear into hate. If something is making us anxious, it's easy to feel negative about it. For our distant ancestors, rumours of an impeding incursion from the people over the river could be quickly fanned into hatred, whether the rumours were true or not. These people could rapidly become designated as worthless and sub-human and a pre-emptive strike justified. Our fearful nature is at the root of tribalism and intolerance.

Hatred of 'the Other', the 'dole bludger', the ethnic minority, the homosexual, the Catholic, the Protestant, the Muslim, the Collingwood or South Sydney supporter, stems from a fear that our beliefs, values or 'way of life' is under threat.

My protagonist in *Influence*, Ziggi Blasko, knows how to turn fear into hate. I felt, however, that it would be a mistake to satirise Ziggi so that he became a joke. He and his ilk are more formidable than that.

So my Ziggi is not just a ruthlessly cynical manipulator. His power over his audience derives from the fact that they sense that he's telling the truth as he sees it. Underneath the confidence and bluster is an often fearful man. And maybe some of his fears are justified. When his sister Connie tells him that his fears of an extremist Muslim attack on Sydney with a 'dirty' atomic bomb are over the top, he replies:

> ZIGGI: If I had've told you on September the tenth 2001 that you could wake up tomorrow with the Twin Towers gone and the Pentagon wrecked you would have said exactly the same thing. 'The probability is tiny.' Well, I'm sorry, but it wasn't.

Ziggi is right to be worried about Muslim extremists. Their fear of Western values has turned into a virulent hatred indeed. And the insensitive behaviour of the world's only superpower towards the feelings, needs and interests of the rest of the world has exacerbated that hatred. But the central question of the play, which must necessarily remain unanswered, is how worried is Ziggi entitled to be?

Influence raises issues, but also, I hope, looks at the eternal human tragi-comedy, as we go about our daily lives. Ziggi might feel he has all the answers Australia needs, but he's got very few answers to the course of his own life. He is human and vulnerable, but that doesn't mean at another level he can't be virulent and destructive. As his father Marko says to his daughter Connie:

> MARKO: Connie, you are *intelligent*, but Ziggi is *smart*.

As every demagogue and dictator knows, being smart, in the sense of intuitively knowing how to manipulate and influence, can be far more potent than mere intelligence.

April 2005

Influence was first produced by Sydney Theatre Company at the Drama Theatre, Sydney Opera House, on 18 March 2005, with the following cast:

ZIGGI	John Waters
CARMELA	Genevieve Hegney
VIVIENNE	Octavia Barron-Martin
TONY	Andrew Tighe
CONNIE	Vanessa Downing
MARKO	Edwin Hodgeman
ZEHRA	Zoe Carides

Director, Bruce Myles
Set Designer, Laurence Eastwood
Costume Designer, Ingrid Weir
Lighting Designer, Peter Neufeld
Composer, Paul Charlier

CHARACTERS

ZIGGI BLASKO

CARMELA BLASKO

VIVIENNE BLASKO

TONY

CONNIE BLASKO

MARKO BLASKO

ZEHRA

ACT ONE

SCENE ONE

Radio studio. Morning.

Lights up on ZIGGI BLASKO, *early fifties. He sits in the broadcast studio of his Sydney radio station. He's a man bursting with opinions he wants to share with his listeners. He waits for his 'easy listening' music selection to finish, tapping his fingers impatiently.*

ZIGGI: That was the great old Eagles standard, 'Fastlane'. And this is Ziggi Blasko on Life 608, the station that lets you get it off your chest. What are your pet hates? I'll tell you one of mine. It's our bleeding-heart, over-educated, left-liberal elites. Forever falling over themselves to tell us how the evil multinational corporations are screwing up our lives. Forget the fact that it's those same corporations who make the laptops the pinot noir set tap out their laments on, or the cars they drive to cushy tenured jobs at our taxpayer-funded universities. And if they're not whining about the corporations that supply their every need, they're ferreting around to find yet more victims of our cruel society. Aborigines, ethnic minorities, gays, women, 'other abled', left-handed lesbians—you name it, they're suffering. The latest bleeding-heart cause is the so called 'working poor'. If your wage is only our current minimum of four hundred and seventy dollars a week, then your life, according to these doozies, is a living hell. The unemployed could be excused for thinking they're doing it tough, but give me a break, someone earning nearly five hundred a week mightn't be in paradise, but they should be able to cope. Sure rents are high in this city, so you can't live in Vaucluse, and you can't drive a BMW and you can't dine out at Tetsuya's three times a week, but with sensible choices and a careful budget, life should be perfectly fine. If you choose to live beyond your means, it's your fault, buddy, no one else's. Hey, the switchboard's lit up. Let's hear what you've got to say out there. Veronica from Ryde.

Veronica, can you credit that these people—the so called 'working poor', are going to charity agencies and brushing aside the real poor and grabbing whatever they can? I don't know about you, Veronica, but I find that disgusting.

VERONICA: [*voice-over*] Ziggi, I don't deny that there are some people doing it tough out there—

ZIGGI: Neither do I. The real poor, not someone who has a job.

VERONICA: [*voice-over*] —but if we don't like our circumstances, we can change them.

ZIGGI: Absolutely. Go to Centrelink. Enrol in a skills course. Get qualified. Don't whine.

VERONICA: [*voice-over*] I was brought up in a family that had next to nothing but I worked my butt off and saved.

ZIGGI: That's a word we don't hear much these days, Veronica. 'Saved'. Now it's put everything on the credit card and whine when they can't pay.

VERONICA: [*voice-over*] I didn't smoke or drink—

ZIGGI: And there's another point. It's a pretty safe bet that most of the so-called 'working poor' are poor because of alcohol and substance abuse.

VERONICA: [*voice-over*] I just knew what I wanted and today I've got it. A beautiful house, successful children and I owe nobody anything.

ZIGGI: Precisely. I'll tell you, Veronica, I didn't grow up with the proverbial silver spoon either. My dad Marko arrived here from war-torn Europe without a cent. And he worked his heart out to provide for us. Whatever I am today I owe to him. He was tough but fair. If we didn't do our share—whack! But he never took off his belt unless we deserved it. Every time I got whacked, I learned something. That's another thing our chattering classes should learn. If the belt came off a few more times these days, we wouldn't have a society full of spoilt young brats, right Veronica?

VERONICA: [*voice-over*] A good smack on the bottom never hurt anyone.

ZIGGI: Veronica, you're the sort of person who brightens up my day.

VERONICA: [*voice-over*] You've brightened up a lot of mine, Ziggi.

ZIGGI: Here's Chicago with 'Hard To Say I'm Sorry'. And I'm not playing this for you, Veronica, because you've no need to be sorry about anything.

*ZIGGI switches himself off air as the music starts and switches
through to his unseen producer.*

Sounded like she might like a smack or two on the bottom herself,
Al. Keep her number.

He gives Al a clenched forearm salute of virility and chuckles.

◆ ◆ ◆ ◆ ◆

SCENE TWO

Ziggi's living room.

VIVIENNE BLASKO, *seventeen, stands in the living room, a suitcase by her
side, looking defiant.* CARMELA, *twenty-nine,* ZIGGI's *second wife, stands
silently on the other side of the room, holding a sleeping baby. The living
room has oodles of space and great art on the walls.* ZIGGI *is very rich.*
ZIGGI *enters and sees* VIVIENNE *and the case.*

ZIGGI: Vivienne?

VIVIENNE: Daddy, I can't stay living with Mum a second longer. It's
just too awful. She screams at me. All the time. I just know if it goes
on I'm going to fail my HSC and never get to uni.

ZIGGI: You want to stay here?

VIVIENNE: Dad, please. I'm desperate.

CARMELA: I've explained to Vivienne that we'd like to help but it's
impossible.

ZIGGI: Carmela, she's my daughter. Maybe I've got a say in this as well.

CARMELA: Ziggi, I wanted a child, fine. It was my choice. But I made it
totally clear that my career is still enormously important to me.

ZIGGI: Having Vivienne here for a little while isn't going to stop you—

CARMELA: For God's sake, Ziggi, do you realise the effort it takes to get
back up on top in my game? The *mental focus* that's required! When
I'm finished at the end of the day this house has to be a *sanctuary*.

VIVIENNE: I won't intrude.

CARMELA: Sweetie, how can you not intrude? You'll be here every
morning, every night.

ZIGGI: Carmela—

CARMELA: Vivienne, I'm not a monster. I'm sympathetic to your plight. After being on the receiving end of many vicious phone calls from your mother, I know what a thoroughly unpleasant person she is. But at this crucial stage of my life my priorities have to take precedence.

ZIGGI: Carmela—

CARMELA: [to VIVIENNE] You think your HSC year is tough. I don't want to sound dismissive, but truly it's a breeze compared with what I'm facing. The ballet scene is a snakepit. A vicious, backbiting, backstabbing, gossip-ridden snakepit full of gormless little nineteen-year-olds will sleep with anyone, male or female, if it helps their career.

ZIGGI: Just let her stay a few weeks till she sorts this thing with her mother.

VIVIENNE: I could help around the place.

CARMELA: Vivienne, you shouldn't promise what you can't deliver. Your father's told me you've been waited on hand and foot all your life.

VIVIENNE: Dad, I've always helped.

ZIGGI: Darling, you have been a little spoilt.

VIVIENNE: It's just that I have to do heaps of study.

CARMELA: Your father said you still don't know how to use a washing machine. The microwave and DVD was your limit.

VIVIENNE: Dad!

ZIGGI: Carmela, all I said was—

VIVIENNE: I'll do my washing. I'll do everyone's washing. Just don't send me back to Mum.

ZIGGI: [to CARMELA] She could be a great help with Freda.

CARMELA: You'd entrust your child to someone who can't even work a washing machine? 'Sorry' doesn't mend the damage when your child's been dropped head-first on the bathroom floor.

VIVIENNE: [to CARMELA] Look, I know I wasn't friendly to you at first, but it was just that Mum was upset and crying every night and I sort of blamed you, because I sort of believed you'd lured Dad away, but I guess I always knew things weren't great between Mum and Dad, and— [that it wasn't all your fault]

CARMELA: Vivienne, your father pursued me. Sent dozens of red roses every day. Couldn't move in my damned apartment for red roses which I've never liked in any case.

VIVIENNE: I know. It wasn't all your fault.

CARMELA: None of it was my fault.

VIVIENNE: And I'm really sorry for the things I said about you to Mum.

CARMELA: Like what?

VIVIENNE: I just said them because I knew it was what she wanted to hear.

CARMELA: Like what?

VIVIENNE: Like that you were totally up yourself.

CARMELA: Oh, lovely.

VIVIENNE: I was just saying it because—

CARMELA: I value my own worth. Do you want women to return to cringing subservience?

VIVIENNE: Look, I know it's going to be hard for you to get back on top. Dad told me that things weren't going too well for you even before the baby—

CARMELA: [to ZIGGI] You what!

ZIGGI: I just said that— [you'd]

CARMELA: I wasn't given a role I deserved because of blatant favouritism. William admits it now. Absolutely admits it.

ZIGGI: All I said was that you'd had a disappointment.

CARMELA: [pointing to VIVIENNE] She's badmouthed me to her mother and she'll do it again. I'm not having a programmed hate missile in my house.

VIVIENNE: I'm not even talking to Mum now.

CARMELA: [to ZIGGI] You might think it's an ego boost that daughter comes home to Daddy, but I'm sorry to tell you, she's here because it's an easier ride for her. More money for clothes and nightclubs and drugs—

ZIGGI: She doesn't take drugs!

CARMELA: You told me yourself you were worried sick about the things she was doing.

VIVIENNE: I'm not taking anything. Just ecstasy.

ZIGGI: Just ecstacy?

VIVIENNE: If you didn't take that you'd be a total dag.

ZIGGI: Well, I'll tell you what, young lady. That's going to stop right now. Carmela, she stays here where I can keep an eye on her. Her mother obviously didn't.

CARMELA: You don't give a damn, do you? William is getting letters and phone calls from ballet lovers all around Australia asking when I'll be back. But you don't give a damn.

ZIGGI: Love, you're being a little hysterical.

CARMELA: Hysterical? I was at the international pinnacle of an art form, Ziggi. Does that mean anything to you?

ZIGGI: I'm just trying to do the right thing.

CARMELA: Obviously not by me.

ZIGGI: By everyone!

CARMELA: I want her out of here as soon as possible.

VIVIENNE: Carmela, I'm not going to be any trouble.

> CARMELA *walks off into the kitchen area.*

ZIGGI: [*calling offstage*] Tony!

> TONY, *a taciturn man in his forties, appears.*

Tony, will you help my daughter with her case? She'll be staying a while. Put it in the upstairs guest room.

> TONY *nods and picks up the case.*

TONY: Do you need me to drive you or Mrs Blasko anywhere tonight, Mr Blasko?

ZIGGI: No, that's all for today, Tony. Pick me up at the usual time tomorrow.

> CARMELA *reappears from the kitchen with an electric jug in her hands.*

CARMELA: Tony, the lawn's dying down near the gazebo.

TONY: Lawn grubs. I'm spraying it tomorrow.

CARMELA: And the hedges really badly need trimming.

TONY: It's on my list, Mrs Blasko.

CARMELA: And the woodwork really needs re-staining.

TONY: That's on my list too.

> TONY *nods and goes upstairs followed by* VIVIENNE *and watched by* CARMELA. VIVIENNE *has a quick look into her room and turns back to* CARMELA, *trying to establish rapport.*

VIVIENNE: [*nervously*] Nice room.

CARMELA: No loud music. No TV blaring. Freda's very sensitive to noise.

> CARMELA *watches them disappear, then turns to* ZIGGI *accusingly.*

ZIGGI: The thing with her mother will blow over. She'll be out of here in a week.

CARMELA: I can't believe you told her my career was on the slide.

ZIGGI: I didn't say that.

CARMELA: I was approaching my peak. It was an absolutely agonising decision to agree to get pregnant.

ZIGGI: You were the one who wanted the baby.

CARMELA: Only because I knew you wouldn't be happy until I did. Next you'll tell me you agree with that low-life arts journalist who hinted that I had the baby because my career was on the skids.

ZIGGI: Of course I don't.

CARMELA: I've kept her article and when I'm given star billing next year I'm going to staple it to the subscription program and send it to her with a sweet little note saying 'Suck, honey, suck'.

> VIVIENNE *comes down the stairs with* TONY, *who nods and moves on out through the kitchen.*

VIVIENNE: [*nervously*] Towels? And soap?

CARMELA: [*sighing and pointing*] The linen cupboard.

> *The front doorbell rings.*

ZIGGI: Who the hell is that?

> *He goes to get it.*

VIVIENNE: I've got some washing. It'd be great if you could show me how to work the machine.

CARMELA: Who do you think I am? The housekeeper?

> *We hear conversation in the hallway and* ZIGGI *enters with his sister* CONNIE, *forty-seven, looking very worried.*

CONNIE: Hi, Vivienne. Ziggi tells me you've moved in.

VIVIENNE: [*embarrassed*] Yeah, er, trouble with Mum.

CONNIE: That's a pity. I like your mum. Hi, Carmela.

CARMELA: [*coolly*] Hi, Connie.

ZIGGI: [*to* CARMELA] There's a problem with Dad.

CARMELA: He's ill?

CONNIE: No, no. He's depressed. In case you hadn't noticed he has been for some time.

ZIGGI: Dad is absolutely one hundred percent fine.

CONNIE: Ziggi, I've been a psychologist for over twenty years now. I know when someone's depressed.

ZIGGI: According to your mob, everyone's suffering from stress or depression.

CONNIE: I know your views on mental health. Whenever I take a taxi I'm forced to listen to them.

ZIGGI: Everyone has stress in their life. Everyone gets down now and then. Instead of whining, cope with it.

CONNIE: Ziggi, our father's depressed.

ZIGGI: He's fine.

CONNIE: Deeply depressed.

ZIGGI: He can't stay here.

CARMELA: Stay here?

CONNIE: [*to* ZIGGI] What am I supposed to do? [*Pointing*] Go back out to the car and tell him his son won't have him?

ZIGGI: He's happy with you.

CONNIE: Not any longer, and frankly, it's time you took some of the load.

ZIGGI: [*pointing*] Vivienne's just arrived.

CONNIE: This place is four times as large as mine.

CARMELA: I'm sorry, Connie, but it's just not possible. Have any of you any idea of just how arduous and demanding my training schedule is?

CONNIE: Have you got any idea how arduous and demanding it is to try and get severely disturbed people functioning again in a totally underfunded government agency?

CARMELA: You can't dump him on us.

CONNIE: He wants to be here. I came home this evening and he had tears streaming out of his eyes because he'd just heard Ziggi telling the whole city what a great father he'd been.

ZIGGI: That doesn't mean that he can live here.

CONNIE: He listens to every minute of every broadcast you make. And agrees with every rotten fascist opinion you utter. [*Imitating her father*] 'Ziggi, he's the voice of the people. They love him because he tells it how it is.' Frankly, I've had enough.

ZIGGI: It's not him driving this, it's obviously you.

CONNIE: It's him! He told me that for the last months of his life he wants to be with his boy.

> MARKO *enters. He's a frail but dignified old man of eighty-two. He stands there frowning. They turn and see him. He moves towards* ZIGGI *and hugs him.*

MARKO: What you said today—it was the most beautiful moment in my life. Suddenly it felt like I deserved my place on earth after all.

ZIGGI: And I, er, meant it, Dad.

MARKO: And now I am so worried that Connie will think I don't love her. It's not that. I love you both. But in my last months on earth I want to be here with you. It's a deep feeling and I can't lie about it.

ZIGGI: Yeah, look, Dad. I'm so touched. Really, but—

MARKO: I listen to you every day and feel like cheering. My son has the guts to take on the incompetent, the stupid, the lazy and the whiners. I want to be here to say 'Bravo' when you come in that door at night.

CARMELA: Marko, it's lovely you feel that way, but the practical problem is that we have a young child, Vivienne has apparently moved in, and I'm about to launch myself back into an international career and I can't be asked to run a boarding house.

CONNIE: [*to* CARMELA] You've got two full-time staff!

CARMELA: [*to* CONNIE] If only it was that easy. I've had to fire two housekeepers in the last three weeks and the four possible replacements I've interviewed this week have been hopeless.

MARKO: [*with dignity*] I understand. I am sorry for being so insensitive.

He turns to go.

CONNIE: Ziggi!

ZIGGI: Dad. You can stay. Of course you can stay.

> MARKO *turns back.*

MARKO: No, Carmela is right.

CARMELA: Marko, I don't want you to think I'm uncaring, but maintaining one's art—

CONNIE: For God's sake, Carmela. The world's not going to fall off its axis if we don't get to revisit your dying swan.

CARMELA: That's right. Anyone will do. Who gives a damn about excellence!

MARKO: I'm not going to impose.

CONNIE: Dad, he's said you can stay. Which is just as well. The press would love to hear that the father he loves so dearly is not welcome in his own house.

ZIGGI: You would, wouldn't you?

CONNIE: Too bloody right. Someone help bring his luggage in.

MARKO: No.

ZIGGI: Dad, it's fine. I'll get your luggage.

CARMELA: I'm not having both of them here.

VIVIENNE: I'll go.

ZIGGI: Vivienne—

VIVIENNE: Don't worry, the Cross is full of homeless kids.

ZIGGI: Vivienne, stay right there.

CARMELA: [*tersely*] It seems my feelings and needs count for precisely nothing. Nothing!

> CARMELA *leaves in high dudgeon.*

MARKO: Ziggi, I can't—

ZIGGI: Dad, it'll be all right. She's just on edge because of this ballet thing.

CONNIE: Stay there. I'll get the cases.

ZIGGI: Vivienne, at least go and help.

> *They hurry out in case* ZIGGI *changes his mind.* MARKO *turns to* ZIGGI.

MARKO: Today was like magic. [*With tears in his eyes*] 'Whatever I am today, I owe to him.' How many fathers would kill to hear those words. Especially when thousands are listening.

ZIGGI: Hundreds of thousands actually. As well as Sydney I go right across regional Australia.

MARKO: I've still got tears in my eyes.

ZIGGI: Dad, I meant what I said. I wouldn't be where I am today without you.

MARKO: I was so moved. Except for that part about the belt. I sometimes threatened to use it but—

ZIGGI: I exaggerated that part a bit for dramatic effect. Sorry.

MARKO: They'll think I'm some kind of sadist.

ZIGGI: Dad, no one's going to think the worse of you for that. I got flooded with calls saying how lucky I was to have a father like you.

MARKO: Yes?

ZIGGI: [*nodding*] Dozens.

> MARKO *looks at his son, then looks at the house around him, and nods happily.*

◆　　◆　　◆　　◆　　◆

SCENE THREE

Ziggi's living room. Some days later.

ZEHRA, *forty-two, a slim woman with her head slightly bowed, faces* CARMELA *who is interviewing her.*

CARMELA: Zehra? How do you spell that?

ZEHRA: Z-E-H-R-A.

CARMELA: What is that? Turkish?

ZEHRA: Yes, Turkish. My mother was Greek, my father Turkish. Not a good mix.

> *She smiles. An appealing smile.*

CARMELA: You've got children?

ZEHRA: Three.

CARMELA: They won't distract you if you work here?

ZEHRA: No, no. They're old enough to look after themselves.

CARMELA: You know how to cope with babies?

ZEHRA: For sure.

CARMELA: We're sometimes going to need you to be here a little longer than eight hours a day.

ZEHRA: What hours would you need me?

CARMELA: I'd like you here by eight and I'd want you to stay on until the early evening to prepare the evening meal and put Freda down.

ZEHRA: I wouldn't like to get home too late. My children.

CARMELA: [*sharply*] I understood they were old enough to look after themselves.

ZEHRA: Oh, yes. Yes. Just—I like to see them.

CARMELA: I'm really looking for someone who can concentrate on this household here, so if you feel that your children need a lot of your attention then perhaps—

ZEHRA: [*anxiously*] No, no. They are fine. It's just I live a long way out and the train ride is very long. And then there's the bus.

CARMELA: Zehra, it's really important that your problems don't become my problems. I have enough to worry about without that.

ZEHRA: I understand. Really. I am here to make things easier for you.

CARMELA: Precisely. Not that I'm uncaring, but much as I'd like to be a sympathetic ear, I'm working extremely hard and I really don't have the energy right now.

ZEHRA: I understand.

CARMELA: Good. Now. Cooking. Ziggi and I tend to eat out a lot so you wouldn't be needed every evening, but the agency did say you were strong in this area.

ZEHRA: My husband and I ran milk bars but then a small restaurant. I did most of the cooking.

CARMELA: Turkish?

ZEHRA: Actually no. I am Turkish, he was Lebanese, so naturally our restaurant was Italian.

CARMELA: I love good Italian.

ZEHRA: I think mine was just— [*She makes a so-so motion with her hands.*] We went broke just before he died.

CARMELA: Are there any questions you have for me?

ZEHRA: Oh, no no. Ah, what is it that keeps you so busy?

CARMELA: Training. I'm a ballet dancer.

ZEHRA: Belly dancer? Hey, that's very Turkish. I did it when I was younger.

    She demonstrates. CARMELA *is horrified.*

CARMELA: Ballet dancer. Ballet. Odette. *Swan Lake.* That's my photo in the hallway.

ZEHRA: [*surprised*] That's you?

CARMELA: Yes, in London.

ZEHRA: You were so beautiful back then.

CARMELA: I hope I haven't deteriorated too much in the interim.

ZEHRA: [*alarmed*] Oh, no no. I meant in that costume. You looked so beautiful. In that costume.

CARMELA: Your references read very well. We'd be offering four hundred and seventy-five dollars a week. Is that satisfactory?

ZEHRA: Oh.

CARMELA: It's the award. Is that a problem?

ZEHRA: No, but the hours…

CARMELA: You won't be required to be on duty more than eight hours. When Freda's asleep and you've finished cleaning and ironing you can watch television.

ZEHRA: Oh, yes.

CARMELA: Is that a problem?

ZEHRA: No. No.

CARMELA: Sunday's off, of course.

ZEHRA: That's fine.

CARMELA: And we really need to start you as soon as possible. Say next Monday?

ZEHRA: Monday.

CARMELA: That's a problem?

ZEHRA: No, that's good. The sooner the better.

CARMELA: Good.

MARKO *wanders into the room.*

Are you looking for something, Marko?

MARKO: [*pointing to the entertainment console*] It's nearly time.

CARMELA: [*irritated*] For what?

MARKO: [*frowning*] For Ziggi's program.

CARMELA: Marko, I don't listen to Ziggi every day. In fact, I very rarely listen at all. I'm not exactly his demographic.

MARKO *looks puzzled.*

[*Explaining*] His audience is poorly educated and barely articulate, and Ziggi plays them perfectly, but it's not for me.

MARKO: Ziggi plays them? He says what he truly feels.

CARMELA: And I totally agree with what he says. The hard-working and talented should not have to subsidise the stupid and lazy, but I don't have to listen to Ziggi to work that out.

MARKO: Never? You never listen?

CARMELA: Marko, I've heard it all, okay? This is—what was your name again?

Edwin Hodgeman as Marko and Genevieve Hegney as Carmela in the 2005 Sydney Theatre Company production. (Photo: Heidrun Löhr)

ZEHRA: Zehra.

CARMELA: Zehra. She's our new housekeeper.

ZEHRA: Pleased to meet you, Mr Blasko.

MARKO: [*ignoring her*] Why did you marry my son if you can't even be bothered listening?

CARMELA: We'll discuss this in just a minute, Marko. I'll just see Zehra out.

She ushers ZEHRA *towards the front door.*

ZEHRA: It's a lovely house, Mrs Blasko.

CARMELA: [*sighing*] Ziggi's taste, I'm afraid. I'd prefer something with grace and history. Victorian or Italianate. Perhaps one day.

ZEHRA stares at her and the house in surprise. She can't conceive that anyone would be dissatisfied with it. They both exit towards the front door.

MARKO goes to the radio, which is part of the entertainment console, and turns on the power.

ZIGGI: [*voice from the radio*] If a kid of thirteen knifes someone, or knocks someone down in a stolen car, then the law should treat them exactly the same as an adult. They've done just the same damage to life and limb as someone of twenty-five, so why pussyfoot around? And I'll go one step further. A large number of these lawless thirteen-year-olds are Aboriginal. And somehow, apparently that means we have to turn a blind eye. And if we dare try and arrest one, our police have to stand there while they're showered with Molotov cocktails. Let's get tough on crime, means all crime, Mr Premier. Not just criminals who are old and white. This is Ziggi Blasko on Life 608, the station that lets you get it off your chest. Let's hear more of what you've got to say about that one, but first, let's hear that easy listening favourite Ol' Blue Eyes, with 'The Lady is a Tramp'.

Frank Sinatra begins to sing. CARMELA *comes back in and turns off the radio. She and* MARKO *glare at each other.*

CARMELA: I'm sorry, Marko, but this is my house. If you must listen I'll get Tony to buy a radio and put it in your room.

MARKO glares at her, gets up and walks off, then turns.

MARKO: Why do you hire a woman like that?

CARMELA: Like what?

MARKO: Arab.

CARMELA: She's Turkish.

MARKO: The same.

CARMELA: I hired her because I think she'll do a good job.

MARKO: They are not quality people.

CARMELA: Marko, your prejudice is incredible. But then again you did fight for Hitler.

MARKO: Against my will. Is she Muslim?

CARMELA: I didn't ask her because I don't give a damn.

MARKO: They hate us.

CARMELA: Sure. She's planning to infect herself with bubonic plague and lick our dinner plates.

>   MARKO *mutters and turns to leave.*

MARKO: I buy a radio myself. I don't want anything from you.

>   *He walks out.* CARMELA *watches him go and shakes her head angrily.*

◆   ◆   ◆   ◆   ◆

SCENE FOUR

Ziggi's living room. Late that night.

MARKO *is sitting in a chair staring straight ahead.* ZIGGI *comes down the stairs in his dressing gown.*

ZIGGI: Dad, I thought I heard someone down here. Why aren't you in bed?

MARKO: I can't sleep. I can never sleep these days. Life is just one long torture.

ZIGGI: We'll get you some pills.

MARKO: They make no difference. Nothing makes a difference.

ZIGGI: You can't sit up here all night.

MARKO: She doesn't even listen to your broadcasts.

ZIGGI: Dad. It doesn't matter.

MARKO: How can it not matter? You are the only man telling this country what it has to hear and she can't even be bothered listening?

ZIGGI: She agrees with it all.

MARKO: Our top politicians beg you to be on your show? And she can't be bothered listening?

ZIGGI: To her, politics is just grubby reality. The arts are all that matters.

MARKO: She danced in London once. When the ballet was on tour. And when their two best dancers were injured.

ZIGGI: [*nodding*] Which was probably her moment of glory.

MARKO: Her career is over.

ZIGGI: She's determined to prove otherwise, it could all be just fantasy. I just can't ever say it.

MARKO: Why not? It's true.

ZIGGI: So is the fact I've got haemorrhoids, but I don't like to hear it. Just go softly with her, Dad. I don't want things to get worse.

MARKO: You told me she loved you. She treats you like dirt.

ZIGGI: I've become an embarrassment to her.

MARKO: Embarrassment?

ZIGGI: I'm honest about what's wrong with this country. Her ballet and opera cognoscenti hate it.

MARKO: Then why the hell did she marry you?

ZIGGI: I was option number two when she broke up with Graeme Bastoni. Dancers have very short professional lives.

MARKO: Bastoni? The guy with billions?

ZIGGI: Yeah. The man who has nothing better to do with his cash than fund fat ladies singing Verdi. And don't the artsy-fartsy crowd love him for it?

MARKO: She was going to marry him?

ZIGGI: Yeah, until she found out he was having a little thing on the side with his yoga teacher and married me to spite him. And she's regretted it ever since. And so apparently has he. They still have little lunches together.

MARKO: What kind of marriage is this?

ZIGGI: The kind that when it breaks up she gets seventy percent of everything I own.

MARKO: This is why you stay together?

ZIGGI: I've worked like a dog for what you see here.

MARKO: It's all about money.

ZIGGI: Everything's about money, Dad. Only I forgot that. I was dragged off to the ballet by some corporate mates and suddenly there's this exquisite creature with legs to die for, doing incredibly sensual things, and I went temporarily insane. Three years later I've got no sex life to speak of and half my assets on the line. We're talking millions of dollars here.

MARKO: You're staying together because of the money?

ZIGGI: Not totally.

MARKO: Why else?

ZIGGI: She's young, photogenic. Good for my public image.

MARKO: There's no love at all?

ZIGGI: [*shaking his head*] You've seen her. She's as self-absorbed as a bear with its head up its bum.

MARKO: I'll go back to your sister's.

ZIGGI: Dad, frankly it might be best.

MARKO: Connie despises me.

ZIGGI: She doesn't despise you.

MARKO: I despise myself.

ZIGGI: Dad. I meant what I said today. You worked your guts out concreting driveways so Connie and I would have a chance. You should be proud.

MARKO: Yeah.

ZIGGI: You should.

MARKO: What about before I came here, eh?

ZIGGI: The Germans conscripted you. You had no choice.

MARKO: Yeah.

ZIGGI: You were a medic. You never fired a shot.

> MARKO *turns away, unable to face his son. There's a pause.*

Dad?

MARKO: When you praised me today, I felt wonderful. For ten seconds. Then terrible.

ZIGGI: Why?

Pause.

MARKO: Because I knew then that before I died I had to tell you the truth.

ZIGGI: You weren't a medic?

MARKO shakes his head and ZIGGI frowns.

What were you?

There's a long pause.

What were you?

Finally it all comes tumbling out.

MARKO: I wasn't conscripted. I volunteered. I literally ran to volunteer. Thousands upon thousands of Croats volunteered. The Germans were our saviours. They'd help rid us of our hated Serbs and crush their cousins, the Russians. The proudest day of my life was when I was allowed to put on a German uniform with its own Croatian insignia. Eighteen years old and suddenly a member of the master race.

ZIGGI: Dad, I don't think I want to hear this.

MARKO: We were sent to Austria to train under the German General Fritz Neidholdt. I worshipped him. We called ourselves 'Vrazja', the 'Devil's Division'. We were headed for the Eastern front to fight the Russians, but then in 1943 it was decided to send us back to our homeland to fight the partisans. If the partisans killed one of us, a hundred Serb hostages were to die. That was the formula. For generations we had dreamed of a Croatia that was a hundred percent Catholic. No Orthodox Serbs. The Germans had given us our chance to make the dream real. Ethnic cleansing? We invented it.

ZIGGI: Dad, I don't want to hear any more. I really don't.

He turns to go, but MARKO pulls him back.

MARKO: Sometimes we didn't waste bullets. We tied them together and threw them off cliffs, we beat them to death with picks and axes—

ZIGGI: [*distressed*] Dad, don't tell me this. Don't.

MARKO: For four hundred years the Serbs had lived amongst us in peace… And it meant nothing. Nothing. We became mad dogs. But eventually you pay. I still see the faces, the eyes. I still hear the screams.

MARKO *looks at his son, who slumps into a chair.*

The Serbs didn't wait around to become hostages. They fled to the hills. And when we finally surrendered to the British they handed us over to them. And they murdered us with just the same ferocity. I only escaped because they were careless. They machine-gunned us then went round and put a bullet in our heads. They didn't bother with me. I must've looked very dead.

MARKO *looks at his son who is sitting there speechless.* MARKO *gets up and walks slowly back towards his downstairs room.* ZIGGI *watches him go.*

◆   ◆   ◆   ◆   ◆

SCENE FIVE

Ziggi's living room. Next week.

ZEHRA *is cleaning, methodically and thoroughly, when* TONY *appears.* ZEHRA *looks up.*

ZEHRA: Hi. You're Tony?

TONY *nods.*

Zehra. I'm the new housekeeper.

TONY: [*nodding*] Kid asleep?

ZEHRA: Yes, lovely baby, isn't she? But then, of course, why wouldn't she be? Her mother is so beautiful.

TONY: Yes.

ZEHRA: Should I make you a coffee?

TONY: No, I'm fine.

ZEHRA: You will pick Mrs Blasko up from her ballet training?

TONY: Yes, in a few hours.

ZEHRA: You'll excuse me if I keep working. I want them to see I'm a good worker.

TONY: Go ahead. I'll leave you.

ZEHRA: No, no. I work and talk. It's fine. I almost didn't get this job.

TONY: Yeah?

ZEHRA: I am so stupid. Every time I open my mouth I say stupid things.

TONY: What did you say?

ZEHRA: I didn't know that photo in the hall was Mrs Blasko and when she told me I said how beautiful she used to look.

> TONY *laughs.*

[*Alarmed*] No, it's not funny. I went home and all that night I went over and over it in my memory to try and see whether I'd been as stupid as I thought, and every time I was.

TONY: I'm sure she'll survive.

ZEHRA: All my life I do the same thing.

TONY: You married? Children?

ZEHRA: Three children. My husband's dead. You?

TONY: Two children. My wife's not dead. Just wish she was.

ZEHRA: No.

TONY: Ran off to North Queensland with some other guy and took the kids. And of course she got custody. And I pay maintenance. I don't mind in one sense. I don't want my kids to suffer, but in another sense I'm as mad as hell. Guy she ran off with was full of bullshit about taking care of her but he can't even get a job. Total no-hoper. My money supports him too. Sorry about your husband.

ZEHRA: Years ago now. Do you see your children?

TONY: Last time was two years back.

ZEHRA: That's not good.

TONY: I burn up inside, but there's nothing much I can do. Costs a fortune to drive up and stay in a bloody motel.

ZEHRA: Australia. Everything's so far.

TONY: Three days on the road.

ZEHRA: How old?

TONY: Two boys, fifteen and thirteen. Young fella looks as if he's going to be a really good tennis player. Older fella's great with computers. I phone them a bit, but that costs a fortune too. Ziggi thinks he's paying me good money but I'll tell you something, he's not.

ZEHRA: Don't talk to me about it. Our rent takes most of it. In the last ten years I've not saved one cent. We eat toast, rice, noodles and sometimes meat.

Andrew Tighe as Tony in the 2005 Sydney Theatre Company production. (Photo: Heidrun Löhr)

TONY: I save just enough for one holiday a year. If I didn't I'd go mad.

ZEHRA: You're lucky. I've had one holiday in my life. And that was my honeymoon. And I was seasick all the time. [*She bursts out laughing.*] Seasick all the time. And we couldn't get off the ship. Theo was so angry.

> *She laughs.* TONY *smiles.*

[*Imitating her late husband*] 'What kind of woman I marry? Bloody sick every day.'

TONY: Last holiday. Gawd.

ZEHRA: Sometimes I feel sorry for myself. Sometimes I cry. But then I think every night I come home and my three wonderful girls are there. And we laugh and I hug them and it's okay. Wouldn't be dead for quids, as they say.

> TONY *looks reflective.* ZEHRA *frowns anxiously.*

See? See? I've done it again. You can't ever see yours and I go chattering on about how wonderful mine are. Oh, I am so, so sorry.

TONY: It's fine.

ZEHRA: Oh God, I can't believe I said that. I just don't think.

TONY: It's fine.

ZEHRA: [*still worried*] I can't believe I said that to you.

TONY: Stop obsessing about it.

ZEHRA: You are kind, but I know you were hurt.

TONY: No, because I know you didn't mean it.

ZEHRA: I'm sorry about your wife. She must be a stupid woman to leave you.

TONY: Not really. Earned next to nothing and spent half of it on booze. She had a right to be fed up. I don't touch the stuff now, but it's too late.

> ZEHRA *nods sympathetically.* VIVIENNE *appears on the stairs, looking ragged, haunted.*

Aren't you supposed to be at school, young miss?

VIVIENNE: Don't call me 'young miss'.

TONY: You're supposed to be at school.

VIVIENNE: I can't face it.

She slumps on a chair and stares straight ahead. TONY *looks at* ZEHRA *and raises his eyebrows.*

[*To* TONY] Tomorrow, could you drive me?

TONY: Mrs Blasko says you have to take the train.

VIVIENNE: Oh, great. Even if she's not using the car I have to take the train.

TONY: I just do what I'm told.

VIVIENNE: I'd love her to have to take the train. Oh, yeah. I'd love that.

ZEHRA: What's wrong with the train?

VIVIENNE: You've seen my school uniform. Could you bear to be seen in it in public? Totally daggy. Totally.

TONY: Better hit the lawn grubs.

He goes. ZEHRA *keeps cleaning.*

ZEHRA: I think you look pretty in it.

VIVIENNE: You'd be in a minority.

ZEHRA: Very pretty.

VIVIENNE: What's your name?

ZEHRA: Zehra.

VIVIENNE: For your information, Zehra, boys think I'm a dog.

ZEHRA: No.

VIVIENNE: They say it. Out loud. And laugh.

ZEHRA: Boys— [*Dismissively*] Pooof! They are the dogs. Always sniffing around. My own Cari, not even thirteen. And she's starting to worry. Will the boys like me? Better if they don't, I tell her. They won't bring you any joy.

VIVIENNE: Are you a Muslim?

ZEHRA: Yes.

VIVIENNE: I wish I was, then I could wear a burka. Why don't you wear one?

ZEHRA: I'm from Turkey. We're not strict like Arabs. Besides, men have stopped looking at me.

VIVIENNE: And me. I've got horrible skin.

ZEHRA: It's good not to be too attractive. You can concentrate on your schoolwork.

VIVIENNE: [*incensed*] Thank you.

ZEHRA: No, sorry. Sorry. Your skin is fine.

VIVIENNE: Yeah, sure.

ZEHRA: What I meant was even if you weren't so pretty, it wouldn't be terrible.

VIVIENNE: A lot of boys *do* think I'm attractive.

ZEHRA: Of course they do. I'm sorry. I think to be safe I just say nothing.

She continues cleaning.

VIVIENNE: Concentrate on schoolwork? Why would I want to? It's dead boring.

ZEHRA: What subjects are you doing?

VIVIENNE: [*reciting them like a litany of horrors*] Commerce, Legal Studies, Geography, English, Society and Culture.

ZEHRA: If I'd had the chance to learn those I wouldn't be doing this.

VIVIENNE: Don't start. You're as bad as Mum.

MARKO comes into the living room. He stares at VIVIENNE.

MARKO: Why aren't you at school?

VIVIENNE: [*with sudden tearful anger*] Because *I hate it!*

She runs up the stairs shedding tears.

MARKO: You'll fail!

VIVIENNE: [*offstage*] Of course I'll fail. That's what I do. Fail. At everything! I'm just a total embarrassment to everyone. I'm just a useless drain on the biosphere! I shouldn't be alive!

MARKO raises his eyebrows, then looks at ZEHRA.

MARKO: I would like to listen to my son.

ZEHRA: Go ahead.

MARKO turns on the radio then as a Fleetwood Mac track comes to a close. He slumps in a chair.

ZIGGI: [*voice-over*] And that was the great Fleetwood Mac. They don't make bands like that anymore. You're tuned into Ziggi Blasco on on Life 608, the station that lets you get it off your chest. Okay, what I'm going to say is going to offend the politically correct, but you know what? I don't care. Because you know what political correctness is to me? It's being able to tell lies—great big whoppers—and get away with it. The Muslim community. 'Oh yes', they all say. 'We're moderates. We're law-abiding and we love Australia.' Really? So how come Australia's Grand Mufti goes to Lebanon and is caught out telling a congregation that September Eleven was God's work. I might be missing something but that doesn't sound moderate to me. Here's my take on this. Until Australian Muslim leaders stop preaching hatred, then our security service should take whatever steps it needs to take to protect decent Australians. Send the ones who are preaching hatred back to where they came from. This country used to be one of the happiest and most secure on earth. Let's try and keep it that way. Look at that board light up. Let's hear what Carl of North Ryde has to say.

CARL: [*voice-over*] Ziggi, thank God you've got the guts to tell it like it is.

ZIGGI: [*voice-over*] Is there any other way?

CARL: [*voice-over*] Muslims have been hating Christians for over a thousand years and it's not going to stop now. I say don't just stop new ones coming. I'm with you. Get rid of the ones who are fouling up our country right now.

> Suddenly ZEHRA *can't take any more. She moves across and switches off the radio.*

MARKO: What do you think you're doing?

ZEHRA: Get rid of us? Did you hear him? Get rid of us?

MARKO: Deport you. Deport those who won't accept that they're Australians.

ZEHRA: Get rid of us?

MARKO: Deport the extremists.

ZEHRA: Where did you come from in Europe?

MARKO: Croatia.

ZEHRA: Hasn't your country seen enough of people who want to 'get rid of' other people?

MARKO: Deport. He meant deport.

ZEHRA: In your country Croat kills Serb. Serb kills Croat. Serb kills Bosnian. Bosnian kills Croat. Where does it all end? Here, I thought. Here. But no, your son wants to bring all that here.

MARKO: He meant deport.

ZEHRA: Words mean one thing then suddenly another. Then it's too late. The world is full of horror, and your son wants to bring it here.

MARKO: Turn that radio back on!

ZEHRA: No. Always I'm saying the wrong thing. And mostly I am sorry. But this time, no. No. No. You tell your son, and he will fire me. But this time I don't care.

> *She stands guard in front of the radio preventing* MARKO *from switching it back on. He finally turns and leaves muttering angrily as he goes.*

MARKO: This is my son's house. And I'm his father. He'll hear about this.

ZEHRA: And he'll get rid of me? Why not? I'm a Muslim. Get rid of us all. Tie our hands and shoot us. Is that what you'd like?

> MARKO *stares at her. The passion and the words she used have impacted on him. The front doorbell rings.* ZEHRA *goes to get it and we hear her introducing herself to* CONNIE. MARKO *walks back towards his room, the radio forgotten.*

CONNIE: [*offstage*] I'm Ziggi's sister. You're the new housekeeper?

ZEHRA: [*offstage*] Zehra.

CONNIE: [*offstage*] I've just come to check on Dad. He seemed very low when I spoke to him last night.

> *They appear.*

Excuse me. I'm just boiling with rage. My car's in for service and I had to catch a cab and, of course, there's my cretin of a brother on the car radio stirring up hatred and the cabbie's saying, 'Yeah, mate, yeah'.

> *Suddenly and impulsively,* ZEHRA *embraces her and bursts into tears.* CONNIE *is taken aback and comforts her as best she can.*

ZEHRA: Sorry, sorry. This is terrible. This is terrible.

CONNIE: You're Muslim?

> ZEHRA *nods.*

[*Pointing to the radio*] You were listening?

ZEHRA: I switched it off.

CONNIE: Good for you. Don't worry about my brother. His brain's only equipped with a right hemisphere.

ZEHRA: I insulted him to your father.

CONNIE: Good.

ZEHRA: He'll tell him, then I'm finished here.

CONNIE: Do you really want to work here?

ZEHRA: It took me weeks to get this job. No one wants to hire a Muslim.

CONNIE: I'll speak to Dad. If he says a word I'll kill him.

ZEHRA: Turkey is secular country. We are not mad extremists.

CONNIE: Zehra, calm down. Not all Australians think like Ziggi.

ZEHRA: There are more like him than you.

CONNIE: No, his side are just louder.

ZEHRA: Was he always like that? Your brother?

CONNIE: [*nodding*] Right from the start he was an aggressive, opinionated little swine. I tried to drown him in the bath when he was four, but he wriggled out of my grasp.

ZEHRA: Here he seems quite nice.

CONNIE: Hitler was very kind to his dogs.

ZEHRA: Why is he like he is?

CONNIE: Bigots are grown-up school bullies. They've worked out that the easiest way to feel powerful is to pick on someone more vulnerable.

    ZEHRA *nods*.

But it's not good to make things easy for them. It was bad that your Mufti called Nine/Eleven God's work.

ZEHRA: [*nodding*] He said it was a bad translation, but I don't think that's true.

CONNIE: It makes Muslims an easy target. Then the more hated you feel the more you hate back. It's a vicious cycle.

ZEHRA: [*nodding*] After Bali, one woman I know, who I thought was moderate, said 'Only eighty Australians dead? What a pity.' How can this stop?

CONNIE: Not by both sides pumping up the hate levels.

ZEHRA: It makes me so worried for my daughters.

She resumes her cleaning.

CONNIE: You look tired.

ZEHRA: No. Around now sometimes a little sleepy, but later I'm fine.

CONNIE: Where do you live?

ZEHRA: Blacktown.

CONNIE: Way out there? You come by train?

ZEHRA: [*nodding*] And bus. It's not so bad.

CONNIE: What time do you leave in the morning?

ZEHRA: Five. But my husband and I ran milk bars. Five is normal.

CONNIE: [*frowning*] When do you get home?

ZEHRA: Nine.

CONNIE: That's a hell of a day.

ZEHRA: My girls are good. I prepare a week's meals on Sunday and they heat them up.

CONNIE: What age are your girls?

ZEHRA: Twelve, ten and six.

CONNIE: I know it's none of my business, but how much does my brother pay you?

ZEHRA: Four hundred and seventy a week.

CONNIE: For working those hours?

ZEHRA: Mrs Blasko says I can rest when the baby's asleep.

CONNIE: You're still here. Still on duty. My brother should pay you more. I'll talk to him.

ZEHRA: [*alarmed*] No, please. I must keep this job.

The sound of rock music is heard upstairs.

CONNIE: Who's that?

ZEHRA: Vivienne.

CONNIE: Why isn't she at school?

ZEHRA: She hates it.

CONNIE: Hates it? Her school fees would keep a thousand African families alive for a year.

ZEHRA: [*puzzled*] She has so much but she is so sad.

CONNIE: Spoilt little brat.

CONNIE *frowns, moves towards the stairs and up them towards the sound.* ZEHRA *watches her go into Vivienne's room.*

◆   ◆   ◆   ◆   ◆

SCENE SIX

Ziggi's living room. Evening.

CARMELA *sits opposite* CONNIE. *They listen as the front door opens and someone approaches. It's* ZIGGI. *He's looking annoyed.*

ZIGGI: I hope this is important, Connie. I had to cut short a very important lunch.

CONNIE: Some corporate giant paying you to give them the odd on-air mention?

ZIGGI: I never say anything I don't believe.

CONNIE: Sure. A mere half million wouldn't sway anyone's judgement.

ZIGGI *sees* ZEHRA *offstage.*

ZIGGI: Zehra, bring me coffee. Strong.

He moves across and gives CARMELA *a peck on the cheek.*

How were classes?

CARMELA: Training, Ziggi. I don't need classes.

ZIGGI: Training.

CARMELA: William came by and he said I'm way ahead of where he expected me to be.

ZIGGI: That's great.

CARMELA: He says Emily and Roseanne haven't lived up to expectations—which I totally predicted—and it's looking good for my early return. [*Looking at* CONNIE] It is very tiring though and I really don't appreciate coming home to this.

CONNIE: Ziggi, I've been with Vivienne most of the afternoon and I'm very concerned.

ZIGGI: About what?

CONNIE: I'd like to have her assessed. I think she's clinically depressed.

ZIGGI: Connie, this is what I can't stand about you. Dad's depressed, Vivienne's depressed. Everyone's a nutcase. It's bullshit. She's fine.

CONNIE: She hasn't been to school for the last four days.

ZIGGI: What?

CARMELA: Great. We feed and clothe her and pay her school fees and she doesn't even bother to go.

ZIGGI: [*moving to the stairs*] We'll see about this.

CONNIE: Be careful with her, Ziggi. Depression accounts for about ninety-five percent of all suicides.

ZIGGI: Don't try that friggin' scare tactic. She's fine. I'll go up there and tell her to get off her arse and get to school or she's back with her mother.

CARMELA: About time.

CONNIE: Ziggi, be careful.

ZIGGI: [*calling up the stairs*] Vivienne!

CONNIE: Don't have it out with her now.

ZIGGI: Like hell I won't.

> VIVIENNE *appears on the stairs.*

VIVIENNE: What?

ZIGGI: Don't say 'What?' to your father.

VIVIENNE: What d'you want me to say?

ZIGGI: Say 'Yes, Dad?', and say it without a surly note in your voice.

VIVIENNE: [*mock sweetly*] Yes, Dad?

ZIGGI: Have you been missing school?

VIVIENNE: Yes! Thanks, Connie. I thought our talk was confidential.

CONNIE: Vivienne, I think you need to see someone.

ZIGGI: What's this stuff about being suicidal?

CONNIE: I didn't say that.

VIVIENNE: Who wouldn't be in this family?

CONNIE: [*to* VIVIENNE] I think you could be depressed.

VIVIENNE: 'Course I'm depressed. I'm doing shit subjects at school, boys call me a dog on the train every day, and I'm living here!

CARMELA: Well, go back to your mother.

VIVIENNE: She's worse.

CARMELA: Your only problem is you're totally indulged.

VIVIENNE: Indulged? That's rich coming from someone who's waited on hand and foot twenty-four-seven!

CARMELA: If I have to put up with one more skerrick of rudeness from you, you can live on the street for all I care!

VIVIENNE: [to ZIGGI *indicating* CARMELA] You want to know why I'm depressed? Try her.

ZIGGI: That is enough, young lady. You're getting the best of everything. Best school, best clothes, best whatever.

> ZEHRA *enters with* ZIGGI's *coffee.*

[*Pointing to* ZEHRA] Zehra's kids don't get everything handed to them on a plate like you do.

> ZEHRA, *embarrassed, waits for* ZIGGI *to take his coffee.*

VIVIENNE: Dad, school is horrible. It totally sucks.

ZIGGI: For every day you miss school you miss a week's pocket money, got it? And don't let me hear any more nonsense about depression. You've got everything you can possibly want.

VIVIENNE: Except someone who cares about me. Or even listens! You don't, Mum doesn't, and she [*pointing to* CARMELA] certainly doesn't.

CONNIE: [*to* ZIGGI] Ziggi, depression is not something you can wish away. It's serious.

CARMELA: [*to* VIVIENNE] I've tried and tried with you, young lady.

CONNIE: [*to* ZIGGI] It's not money kids want. It's attention and love.

VIVIENNE: [*to* CARMELA] Tried? Are you joking? I've grovelled and you still treat me like dirt.

ZIGGI: [*to* CONNIE] She gets attention. She gets love. All they want at her age is to make it into the 'cool' group at school and have boys ring up every night. [*To* VIVIENNE] Everything'd work out if you occasionally put a smile on your face.

CONNIE: [*to* ZIGGI] Smile? There's a profound new approach to mental illness.

ZIGGI: Better than your psychobabble and a hell of a lot cheaper.

VIVIENNE: [*to* ZIGGI] Smile? Are you serious? What's there to smile about here?

CONNIE: Fine, Ziggi. Ignore everything I tell you. But Vivienne has problems and Dad is in deep depression. He hardly ever comes out of his room.

ZIGGI: Connie, Vivienne's fine. Dad is fine. If you've got nothing else to say, maybe you could leave us in peace.

ZIGGI *takes his coffee from* ZEHRA.

CONNIE: [*looking at* ZEHRA] I have got something else to say actually. That diatribe about Muslims today was absolutely disgusting.

ZIGGI: The spiritual leader of Australia's Muslims said that September Eleventh was the work of God and I'm supposed to say nothing?

CONNIE: He speaks for a tiny minority.

ZIGGI: Connie, he's their spiritual frigging leader.

CONNIE: He speaks for a tiny minority. Right, Zehra?

ZEHRA: [*embarrassed*] Yes. Coffee anyone?

CONNIE: For Muslims like Zehra all their energy goes into feeding and caring for their families. Do you honestly think she's about to strap a bomb around her waist?

ZIGGI: I wasn't talking about Zehra.

CONNIE: Well, to Zehra it sounded like you were. Muslim extremists are a tiny minority.

ZIGGI: It only takes one extremist to arrive at Sydney Airport self-infected with the Ebola virus and half the city's population could be wiped out.

CONNIE: Don't be so alarmist.

ZIGGI: There's already enough nuclear material unaccounted for to make a hundred dirty bombs. This city uninhabitable for hundreds of years doesn't alarm you a little?

CONNIE: The probability is tiny.

ZIGGI: If I had've told you on September the tenth 2001 that you could wake up tomorrow with the Twin Towers gone and the Pentagon wrecked you would have said exactly the same thing. 'The probability is tiny.' Well, I'm sorry, but it wasn't. And because John Howard has had the guts to stand side-by-side with President Bush we're a prime target.

CONNIE: That part I believe. Ziggi, hate mongering will only increase the risk.

ZIGGI: Since when has the truth been hate mongering?

CONNIE: What you're doing is making things worse! I came home and Zehra's in tears.

ZIGGI: This is a clash of civilisations. A fight to the death between free, secular democracies and medieval, clerical dictatorships. Idiots like you think a little love, understanding and tolerance will see us through. I'm sorry that you're caught in the crossfire, Zehra, but the whole world is polarising and you'll have to take sides.

CONNIE: What are you suggesting we do? Nuke the whole Muslim world?

ZIGGI: The West has a right to use military intervention to force them out of the Stone Age.

CONNIE: More military intervention? Are you crazy?

ZIGGI: You'd rather wait until the terrorists have portable nuclear weapons? Wake up one morning and Sydney's gone? We have to take on the Muslim religion head-on, before it's too late.

CONNIE: Most Muslims are just like Zehra. All that's on their minds is the struggle to make a new life in a new land. Go and look in the public parks at the weekend. They're there with their families. Eating, laughing.

ZIGGI: Assimilation works with other cultures, but not with Muslims.

CONNIE: Ziggi!

ZIGGI: Just last year second and third generation Muslim kids in London burnt the British flag and swore allegiance to Osama Bin Laden. In perfect English accents on peak-hour television.

CONNIE: A tiny minority.

ZIGGI: Hundreds of them. Shouting, 'All Jews must die'.

CARMELA: The truth is that if Israel was put back in its bloody place the problem would disappear overnight.

ZIGGI: Carmela, Israel is a democracy in a sea of fundamentalism.

CARMELA: Ziggi, the mad Jewish fundamentalists are causing this problem. Building more and more settlements on West Bank Arab soil because God tells them to do it.

CONNIE: I have to say you're absolutely right on that one.

ZIGGI: Oh! The far left and far right combine to serve up virulent anti-Semitism.

CONNIE: You can't silence all debate on the role of Israel by screaming anti-Semitism.

CARMELA: The Jews came back after two thousand years and stole Arab land and they're still doing it. And America, which is run by bloody Jews, supports them to the hilt.

CONNIE: I can't quite go along with that.

ZIGGI: It's just crude anti-Semitism.

CARMELA: Have you ever asked yourself *why* there's so much anti-Semitism? Jews have refused to assimilate into any society they've ever been given refuge in.

CONNIE: Carmela, people have a right to retain their identity.

CARMELA: Identity? Cling onto idiot customs and rituals that make absolutely no sense. Don't eat pork and don't eat shellfish. Why? Because it's part of our 'identity'. What utter rubbish. This is one world. With one set of rational civilised values. In my profession it doesn't matter whether a dancer is from Kazakhstan or Carolina. The truth is we'd have world peace in ten minutes if America invaded Israel and confiscated their weapons of mass destruction, which they actually do have.

ZIGGI: That's brilliant geopolitics that is.

CONNIE: She's partly right.

ZIGGI: Here we go. Far left, far right in the dance of death.

CARMELA: You can browbeat your stupid listeners on air, Ziggi, but you can't brainwash me.

ZIGGI: No, because you don't happen to have a brain.

VIVIENNE: It's not Israel's fault. Our teacher says it's Western freedom and modernity the Arabs hate.

CARMELA: Your Jewish teacher?

VIVIENNE: She's right.

CARMELA: Perhaps you'd better wait until you have a little more experience of life before you throw your opinions at us.

ZIGGI: Carmela, she's making a perfectly valid point.

CARMELA: That's right. Defend her. Little darling can say or do nothing wrong.

CONNIE: It's all about America securing oil resources.

ZIGGI: [*to* CARMELA] I've just disciplined her for missing school, for God's sake.

CARMELA: I want her out of here.

VIVIENNE: Your money didn't pay for this place. Dad's did!

CARMELA: I have borne your father a child. Don't you dare speak to me like that. [*Pointing to* VIVIENNE] Either she goes, or I go.

ZIGGI: Carmela, don't do this. [*To* CONNIE] It's not about oil. It's about democracy.

CONNIE: If America and Israel weren't behaving like the world's bully boys there wouldn't be a problem.

ZIGGI: Israel and the US are democracies!

> ZEHRA *takes his empty coffee cup.*

Zehra, I'm sorry. It's going to get ugly. And your community has to accept a lot of the blame.

> MARKO *comes out of his room and stays in the background, unnoticed by everyone.*

CONNIE: So what's Zehra supposed to do? Find another country?

ZEHRA: [*bursting out*] I don't want another country. I am Australian. I have citizenship. They gave me a certificate. They gave me a gum tree. The tree is now thirty feet tall. I am Australian. My children are Australian. I never listen to the bloody Mufti. He's as crazy as you are!

ZIGGI: Zehra, I'm just telling it like it is.

ZEHRA: 'Get rid of Muslims.' That's what you were saying. 'Get rid of them.'

ZIGGI: Deport fanatics, that's all I was saying. Deport fanatics.

ZEHRA: 'Get rid of'. Those were your words.

ZIGGI: Sorry. Those words were never used.

CONNIE: Ziggi, they were!

> MARKO *speaks some words loudly in Croatian. They wheel round to face him.*

ZIGGI: What?

MARKO: 'Get rid of them' in Croatian.

ZIGGI: Dad!

> MARKO *says the Croatian phrase again.*

MARKO: They are not good words, Ziggi. I've told you where I've heard them before.

ZIGGI: [*alarmed*] Dad, don't go there.

CONNIE: Go where?

MARKO: They told me 'Get rid of them', and I did.

CONNIE: When?

MARKO: In the war.

ZIGGI: Dad, not here. Not now.

MARKO: Yes, now.

CONNIE: Dad, you were a medic.

MARKO: Ask your brother what I was.

ZIGGI: Dad, just shut up. If the media finds out what you did my career's over.

CARMELA: What did he do?

From left: Edwin Hodgeman as Marko, Zoe Carides as Zehra and Vanessa Downing as Connie in the 2005 Sydney Theatre Company production. (Photo: Heidrun Löhr)

MARKO: I fought for the Nazis. I killed Serb hostages.

CARMELA: [*to* ZIGGI] Good God. Get him out of here.

CONNIE: Dad, you couldn't have.

ZIGGI: Dad, I'm just about to start negotiating my new friggin' contract.

CARMELA: What do you think it's going to do to my career?

ZIGGI: Your career's peanuts. This will be the most lucrative media deal in Australian broadcast history!

CARMELA: [*outraged*] Peanuts. My career peanuts?

ZIGGI: In money terms.

CARMELA: Art is not measured in money terms.

CONNIE: Dad, you can't be serious?

ZIGGI: The proudest day of his life apparently was when he put on a Nazi uniform.

CARMELA: Oh migod. Imagine if that got into the press.

VIVIENNE: Grandad, you couldn't have.

CONNIE: Dad, when people have been through a war they get traumatised. Shells explode nearby. Stories they've heard become their own memories. [*To* ZIGGI] The sooner we get him into therapy the better.

ZIGGI: Therapy? And spill his guts to someone else?

CONNIE: If he believes he'd done terrible things he might kill himself.

ZIGGI: Connie, I'm about to go and buy a length of rope and leave it in his room.

VIVIENNE: Dad, that's horrible!

CONNIE: Ziggi, if there's one thing I'm absolutely certain of, it's that our father couldn't kill another human being. He'd threaten to give us a belting but never did. Remember when our cat had kittens? Mum told him he had to drown them and he drove miles to take them to the RSPCA in the hope someone would take them as pets. There's no way he could kill.

MARKO: The kittens were not my enemy.

CONNIE: Dad, there's this thing called false memory—

MARKO: My memory is perfect. Too perfect. You have no idea. Any of you.

CONNIE: Of what?

MARKO: Of what any of us are capable of doing.

CONNIE: Dad, you're imagining this.

MARKO: I wish I was. We were given an option. Of over two hundred men in our company, one hundred and eighty-three volunteered to kill hostages. Innocent hostages.

CARMELA: Then you're animals!

MARKO: [*angry*] We're all animals. When your heart is full of hate we're all animals.

CARMELA: For God's sake, don't try and excuse yourself. That only makes it more obscene.

CONNIE: Dad, I grew up with you. You're not capable of it.

MARKO: Connie, look at me. I killed innocent men. And I've lived with the memories every day since.

She stares at him, then looks away.

VIVIENNE: Grandpa, why?

MARKO: We were patriots.

CARMELA: Patriots? Killers.

MARKO: We loved our country and hated its enemies.

CARMELA: You were killers.

MARKO: Yes! And proud of it. Why do you think it was easy for those Muslims to fly into the Twin Towers? They were full of hate, like I was.

ZIGGI: Dad. That's no excuse. Civilised people control their emotions. I hate extremist Muslims, but I don't kill them.

ZEHRA: [*unable to stop herself*] You kill them with words!

There's a silence.

I'm sorry.

She turns, distressed, and goes back to the kitchen.

CONNIE: Dad, if you did…

MARKO: I did.

CONNIE: You should give yourself up.

ZIGGI & CARMELA: [*together*] Are you crazy?

VIVIENNE: Connie, they'll put him in prison. He's too old.

ZIGGI: [*to* CONNIE] You want the world to know?

CARMELA: Connie, for God's sake.

CONNIE: Dad, it's why you're depressed. It's the only way you're going to get any closure and peace.

ZIGGI: Connie, just shut the fuck up! My whole future could go up in flames.

CONNIE: Ziggi, this is not about you.

ZIGGI: It's about the end of my career.

CONNIE: Every cloud has a silver lining.

MARKO: [*to* CONNIE] I want to confess. Believe me I want to. It's like an ache inside me. But I can't. I can't do that to you and Ziggi and Vivienne. The only thing that keeps me sane is that my hard work has made a better life for you. What I did is not your fault and I can't stain you all. My family is all I have. I can't allow my sins to destroy you.

There's a silence.

I can't allow my sins to destroy you.

He turns and goes back to his room. They watch him go. They look at each other.

CONNIE: He'll have to confess.

ZIGGI: Connie, just lay off!

END OF ACT ONE

ACT TWO

SCENE SEVEN

Radio studio. Monday morning. A week later.

Lights up on ZIGGI *waiting for his 'easy listening' selection to finish, tapping his fingers impatiently.*

ZIGGI: That was the wonderful John Denver with 'Rocky Mountain High'. And this is Ziggi Blasko on Life 608, the station that lets you get it off your chest. Violence. It's something I abhor. And so do ninety-nine percent of you out there. So how is it that after every weekend a report lands on my desk that makes this city sound like a war zone? A man dragged out of his car in Bellevue Hill and beaten half to death and his Mercedes stolen. In his own driveway. Three more car jackings reported, two with women being beaten in the process. Five men stabbed in a gang brawl at Star Casino under the noses of so-called 'security guards'. Two men in balaclavas shot another man dead in front of a hundred witnesses on a crowded street. Another drive-by shooting wounds three. And what do all of these have in common? I'm not supposed to say, am I? The words 'Lebanese', 'Muslim', 'gangs' and 'South West Sydney' are never supposed to pass my lips. The culture of murder, gangs, knives, car 'rebranding', intimidation and sheer total thuggery just appeared out of the blue sky according to the wimpish legions of the politically correct, who still seem to think they can tell us all what we can say and can't say. Stan from Turramurra.

STAN: [*voice-over*] I get really angry, Ziggi, at the pressure being put on police and reporters not to name the ethnic groups involved. For God's sake, let's call a spade a spade. We're entitled to know the dangers facing us out there and who's responsible.

ZIGGI: Thanks, Stan. Brigitte from Lewisham.

BRIGITTE: [*voice-over*] Ziggi, there are good Lebanese Muslims. There are law-abiding Lebanese Muslims, but there are also plenty who aren't. So why doesn't our Premier take the shackles off our police force and let them get in there and do what has to be done to return this city to one in which we can step outside our front doors without fear of being shot?

ZIGGI: Thank you, Brigitte. And so say all of us.

◆   ◆   ◆   ◆   ◆

SCENE EIGHT

Ziggi's living room. The same morning.

CONNIE *sits there with* MARKO. *He's looking away, trying to avoid her eyes.*

CONNIE: Dad, just tell me what you did. Exactly.

>   MARKO *says nothing.*

I have to know.

MARKO: Why? So your nightmares can be as bad as mine.

CONNIE: I'm your daughter. Your flesh and blood. Whatever you did affects me. How many did you kill?

MARKO: Two.

CONNIE: How?

MARKO: Bullets. They were made to lie face down. We shot them in the back of the head.

CONNIE: How could you do it?

MARKO: I hated them.

CONNIE: Dad, you didn't even know them.

MARKO: [*suddenly angry*] You've never been in a war. You have no idea. I'd seen three boys I'd grown up with blown into blood and gristle by a booby trap. Something clicks and you're not human anymore. The second one jerked his head round in fear and looked up at me. And I spat down at his face and blew his head apart. But then when the blood and mud and the pain are over, you become human again. I still see his eyes. They come back and back.

>   *There's a silence.*

When we were told to save bullets and club them to death, I couldn't do it. I could shoot, but I could not bear to crack a human skull. And I was teased. Can you believe that? Teased because I couldn't smash a human skull. That's the full horror of war. That a man can feel deeply ashamed because he can't crack human skulls.

CONNIE: Women and children? Were they killed?

MARKO: No.

CONNIE: Rape?

MARKO: [*angry*] In all wars, always.

CONNIE: You?

MARKO: No, not me. Believe me, not me.

CONNIE: Did Mum know about this?

MARKO: [*shaking his head*] I thought if she ever found out she might stop loving me.

> CONNIE *is silent.*

So now you think, 'Oh, my father is a monster. He is vile. His blood is tainting my veins.'

CONNIE: Dad, no.

MARKO: Read history. From the beginnings of time it is one long horror story. Ghengis Khan, Attila the Hun, the Vandals, the Goths, the Romans, the Barbarians, the Turks—slaughter of the innocent was routine. Routine. Women and children and babies. A city surrendered and the slaughter and rape began. Routine. I have read all about these things. The Germans didn't invent evil. They just used technology to speed things up. Even here. All it took were a few speared sheep, and farmers hunted and shot men, women and children right here on this multi-million dollar foreshore.

> CONNIE *is silent.*

I loved your mother. I was a loving father to you and Ziggi. I've held you both in my arms and almost melted with tenderness. Many times. Many times. When there is work and food and peace and friends there's no reason for fear and hate. It's easy.

CONNIE: Dad, unless you confess, this will haunt you for the rest of your days.

MARKO: You want to be the daughter of a war criminal?

CONNIE: There were men, men that you knew, who in the same situation wouldn't kill, weren't there?

MARKO: Some.

CONNIE: Then saying that everybody would've done it just isn't true.

MARKO: It will finish Ziggi's career.

CONNIE: Dad, Ziggi is a sad, sick human being doing huge social harm.

MARKO: It will stain your life.

CONNIE: It'll make me proud that you had the guts to do it.

MARKO: Connie. I can't.

> MARKO *gets up and walks towards his room. She sighs and gets up to go. The front door opens.* CONNIE *looks up.* CARMELA *enters with* TONY, *looking as if she's had a hard day. She sees* CONNIE *and frowns.* TONY *moves on out to the kitchen area.*

CONNIE: I came to see Dad.

CARMELA: It'll be absolutely disastrous for Ziggi and me if he confesses.

CONNIE: You don't think I'm not going to be tainted?

CARMELA: We're high profile, for God's sake!

CONNIE: Oh migod!

CARMELA: You might enjoy being related to a mass murderer, but I don't.

CONNIE: He killed two.

CARMELA: Is that supposed to make me feel better?

CONNIE: I don't think that quite qualifies as a 'mass' murderer.

CARMELA: If it doesn't worry you, take him.

CONNIE: It worries me a lot.

CARMELA: Ziggi's offered to pay for him to go into a home, but the old bastard won't go.

CONNIE: He needs support.

CARMELA: Who cares what *he* needs, for God's sake?

CONNIE: I do.

CARMELA: The truth is, you're totally determined to destroy us.

CONNIE: Not really, but it is a pleasant thought.

> CONNIE *walks towards the front door.*

CARMELA: You are so pukingly self-righteous.

> CONNIE *walks out without answering.*

◆   ◆   ◆   ◆   ◆

SCENE NINE

Ziggi's living room. Three weeks later. Afternoon.

CARMELA: Zehra!

> ZEHRA *comes in from the kitchen.*

How's Freda been?

ZEHRA: A little upset earlier on. Teeth I think. But she's asleep now. How was your day?

CARMELA: Exhausting. Absolutely exhausting. Sometimes I feel like giving up, but I'm just a whisker from my absolute peak, so I'm sticking in there. And, frankly, I'll be needed. The supposed new talent hasn't lived up to expectations. Typical scenario. Overpraised when they first appeared, went to their heads, didn't work hard enough and now—phhhh! Some critics still sing their praises because they don't want to admit they made a huge mistake, but anyone who knows anything in the profession knows just how mediocre they are.

ZEHRA: That's sad.

CARMELA: The arts is obsessed with hot, new talent. Then they learn the hard way.

ZEHRA: Mrs Blasko, is it possible I can go early today?

> CARMELA *looks at her.* ZEHRA *almost can't go on.*

My oldest girl is part of the school choir. They're doing a performance this evening.

CARMELA: I don't like this sort of thing, Zehra. First it's a school concert, then they've got toothache, then they need help with their homework and before you know where you are, you're never here.

ZEHRA: No, no. This is just very special. She has a wonderful voice.

CARMELA: Yes, I'm sure, but—

ZEHRA: You have such a love of the arts that I thought you'd understand. They're singing all the great songs of Andrew Lloyd Webber.

CARMELA: Zehra, people who know anything don't mention the words 'art' and Andrew Lloyd Webber in the same breath.

ZEHRA: I don't know the songs. We don't have a record player. But my daughter says they are very beautiful.

CARMELA: Then I can only say you should get your daughter out of that school and get her enrolled in a decent one as soon as possible. No, I'm sorry, Zehra. I can't let this sort of thing start to happen. What your responsibilities are outside of here are your business and I'm sorry, but I really don't want to know about them. You're here to focus on my needs and Freda's needs. It's not that I don't care, it's just that I'm facing a challenge much tougher than any you'll ever face, and I really can't afford to be distracted. Do you understand?

ZEHRA: [*with dignity*] I'm sorry I asked.

CARMELA: I mean, if one of your children was very ill, of course I'd let you go, I'm not a monster.

> TONY *comes in from the other room.*

TONY: Sorry, Mrs Blasko, but when I had my kids, the school concert was something… special. They're so hurt if you don't make it. Could I maybe go and get takeaway for you all tonight. That Indian place you like?

CARMELA: Oh, sure. Just add another seventy dollars or so to housekeeping. Tony, I don't really appreciate your interfering. What happens to the food that's been brought for tonight? Just goes to waste?

ZEHRA: No, it'll be fine in the fridge.

CARMELA: That does bring me to something else I have to raise. Zehra, I've been checking the outgoings and the food bills seem really high. Are you taking some home for your family?

ZEHRA: [*hanging her head*] When there are leftovers. Or food that would spoil.

CARMELA: You're not by any chance buying so much that some of it's bound to spoil?

ZEHRA: No.

TONY: It's my fault, Mrs Blasko. I said it was stupid to waste it, and it was only the odd scrap.

Genevieve Hegney as Carmela in the 2005 Sydney Theatre Company production. (Photo: Heidrun Löhr)

CARMELA: There's a bit more than the odd 'scrap' going out this door. What is this? A below-stairs revolt?

ZEHRA: I'm sorry, Mrs Blasko. It won't happen again.

CARMELA: Frankly, it'd better not.

TONY: What little was going wasn't hurting. Zehra can barely make ends meet.

CARMELA: Tony, that is just so ridiculous! Nearly five hundred dollars a week. Are you joking?

ZEHRA: Tony, please don't—

CARMELA: [*an angry outburst*] Look, this is all making me enormously distressed. You two have no idea of the ordeal I'm going through and the strain it's putting me under. I am attempting the artistic equivalent of climbing Mount Everest in thongs, and I come home to this! A school concert may loom large in your psyches, but good God, it's something that I shouldn't have to even hear about!

ZEHRA: I'm sorry, Mrs Blasko.

CARMELA: [*still highly emotional*] Go early, for God's sake. Go and hear Andrew Lloyd Webber. And ponder what constitutes fame in the kind of shrill, shallow society we've become. Get the Indian food, Tony. And see if you can both have enough consideration for what I'm going through at the moment to not upset me with this sort of thing again!

She storms off up the stairs then turns back.

[*To* TONY] Don't get that vile vindaloo!

She continues on up. ZEHRA *is upset.*

ZEHRA: I can't go early.

TONY: No, you go.

ZEHRA *shakes her head.*

No, you go. You can't miss something like that.

ZEHRA *looks at him.*

You go. I'll drive you to the station.

ZEHRA *looks at him and nods.*

◆   ◆   ◆   ◆   ◆

SCENE TEN

Ziggi's living room. That evening.

The door opens in the hall. ZIGGI *enters looking a little jaded.* VIVIENNE, *who has heard her father coming, bounds down the stairs. She's excited and talks breathlessly. The words come in unstoppable surges.*

VIVIENNE: Dad, Dad.

ZIGGI: Yeah, honey.

VIVIENNE: I thought about what you said. There is a fight between Democracy and Mediaevalism and Democracy's only going to win if our economies are strong enough to defend our values.

ZIGGI: Absolutely.

VIVIENNE: As soon as I told myself I wanted to join the fight I started to feel great. It was like I'd been under a whole fog of gloom and suddenly it lifted.

ZIGGI: Yeah?

VIVIENNE: I worked out what I want to be. An investment analyst. It's fantastic. You just work out all these parameters and see what stacks up. Remember how I used to love playing monopoly? It's like that, only better. If you do it right you can't lose. If you—

ZIGGI: You can always lose.

VIVIENNE: No, it's foolproof. You start with price-to-earnings ratio, then you look at management's track record, then—oh, dozens of things, and it's really, really exciting. We learned how to invest on the internet.

ZIGGI: Actual trading?

VIVIENNE: Simulated. I wish it was real. I would've made two thousand dollars.

ZIGGI: You sound like you really are liking school?

VIVIENNE: I'm loving it, I'm loving it. I was reading for my Society and Culture assignment and it talked about self-stereotyping, and I realised, hey, of course, that's what I was doing. I was telling myself that I was stupid, and everything was being filtered through that perception, so of course everything I read didn't make sense, because

how could it, eh? I was stupid. But when you smash that perception—
when you ask yourself what evidence there is for that self-
perception—and you realise—hey, there's none. I always did well
before this year so—hey—how the fuck could I be stupid—?

ZIGGI: Don't swear, honey. It cheapens you.

VIVIENNE: So you abandon that self-perception and suddenly, hey, you're
smart, and the books that made no sense are suddenly the easiest
things in the world to understand. There's a world there inviting
you to conquer its secrets, and migod, what secrets. I thought I looked
terrible and boys hated me, but again—where's the evidence? No
evidence. Smash that stupid bit of self-stereotyping and suddenly
they're all over you like a rash—

ZIGGI: Honey, you're too young for that—

VIVIENNE: Dad, I'm not going to waste my time on the pimply little
scabs at my school. The brilliant thing is you realise that once you've
ditched all the negative self-stereotyping you can tackle anything.
Life is really, really, really there for the taking and I'm taking. You

*John Waters as Ziggi and Octavia Barron-Martin as Vivienne in the
2005 Sydney Theatre Company production. (Photo: Heidrun Löhr)*

see the patterns, you see the possibilities, and everything, everything I'm studying makes sense as a total pattern of preparation. One of my teachers said today that she can't believe the transformation. She said that I was on my way to becoming something very special and that she was going to be very proud that she'd taught me one day. And— [she also said]

ZIGGI: I'd love to hear more, honey, but I'm really tired.

VIVIENNE: Research shows that people are rarely tired in a physiological sense. We need far less sleep than we think we do. It's just another case of self-stereotyping—i.e. 'I'm the kind of person who gets tired'—so, of course, what happens? You *do* get tired.

ZIGGI: I get tired because I am tired! My ratings have started dipping for the first time in years. Just when the contract is up for renewal. They say I'm losing my bite.

VIVIENNE: Don't talk yourself down, Dad. It's all a matter of self-perception.

ZIGGI: It's about my ratings dropping and it's about the fact I can't go on being a money-making machine for you lot forever! I'm fifty-bloody-three!

She follows him up the stairs, chattering at high speed all the way.

VIVIENNE: Dad, with the capital you've got you could make yourself a fortune. Never worry about money again. If you just sit down with me I'll show you how to invest on the internet.

ZIGGI: I'm sure you could, honey, but I'm just a little tired.

He moves off to his room and shuts the door. VIVIENNE, *concerned, watches her father go.*

◆　　◆　　◆　　◆　　◆

SCENE ELEVEN

Radio studio. Next day.

ZIGGI *waits for his 'easy listening' selection to finish, as always, tapping his fingers impatiently.*

ZIGGI: That was Billy Joel with one of his greatest. And this is Ziggi
    Blasko on Life 608, the station that lets you get it off your chest.
    Well, for years I've suspected that the mental health industry was
    one big con. Well, let's qualify that. There are people out there who
    have mental problems, and severe ones, and they do deserve help.
    But all this rubbish about stress and depression is just a way to keep
    thousands of so-called mental health gurus making a very good living
    by encouraging perfectly normal people to believe that there's
    something wrong with them. Stress—we all have stress in our lives.
    So what's new? My sister's in the counselling industry and she told
    me my daughter needed help. She was 'depressed'. She needed
    therapy. And drugs. I told her to get lost. And thank the Lord I did. A
    few weeks later and my daughter has totally got over her perfectly
    normal dip, and she's fine. In fact she's better than fine. She's firing
    on all cylinders. Let's hear what you think. Alan from Bilgola.
ALAN: [*voice-over*] All this mental health psychobabble is just a modern
    form of witchcraft, Ziggi.
ZIGGI: Precisely, Alan. And if I had've listened to the witch doctors, my
    kid would've been zombified out on some barely-tested drug. Gussie
    from Lakemba.
GUSSIE: [*voice-over*] Ziggi, I think the drug companies have invented
    this so-called epidemic of depression to drum up more business.
ZIGGI: Gussie, there's an awful lot of rubbish talked about drug
    companies. I had lunch with some top executives recently and I was
    impressed by how honest and transparent these guys were. The real
    culprit of this scenario is our taxpayer-funded caring industry. This
    is Ziggi Blasco and look how the panel's lit up over this one.

◆   ◆   ◆   ◆   ◆

SCENE TWELVE

Ziggi's living room. Six weeks later.

ZEHRA *is cleaning in the living room. The front door opens and* ZEHRA
starts nervously. CARMELA *and* TONY *come into the room.* CARMELA *is*

looking elated. TONY *moves on quickly into the kitchen giving* ZEHRA *a glance as he goes.*

CARMELA: Zehra, don't fuss around in here. Have you done the offices and the cabana?

> ZEHRA *stares at her, frozen.*

Did Freda have her medicines? In the right order? I'm very worried about that cough of hers.

> ZEHRA *is still unable to answer.*

Zehra, are you listening to me. This room is fine. Have you done the offices and the cabana?

ZEHRA: I have been so worried.

CARMELA: [*alarmed*] Has Freda's cough got worse?

ZEHRA: [*shaking her head vigorously*] About what I told you before you went. Why I was late.

CARMELA: You were only ten minutes late. I was cross, but for heaven's sake, these things happen. I was edgy about today. William was coming in to have a look at progress. Which he did and the verdict? 'Fantastic'. One word, that said it all. 'Fantastic'. I'm next to certain that the phone will ring tomorrow and he'll ask me back as principal dancer.

ZEHRA: I shouldn't have said anything about the accident. I'm so sorry.

CARMELA: Look, you were only ten minutes or so late. I'm sorry I was cross.

ZEHRA: You don't want to hear those things. I know. I'm sorry.

CARMELA: No, it's fine. How is your little girl?

ZEHRA: Upset, but she'll be fine.

CARMELA: The scarring?

ZEHRA: [*worried*] I told you? About the scarring?

CARMELA: Yes, you did.

ZEHRA: I'm sorry. I'm sorry. I didn't mean to talk about it.

CARMELA: Zehra, calm down. It's not a big issue. Have you done the offices and cabana? Look, forget the offices and cabana. Just sit down.

> ZEHRA *says nothing.*

Zehra, forget them.

ZEHRA *says nothing.*

Zehra?

ZEHRA: I told you about the scarring?

CARMELA: Yes, and you shouldn't feel guilty if you did. I'm not a monster.

ZEHRA: You don't want to hear these things. I know.

CARMELA: Zehra.

ZEHRA *stands there immobilised.*

Zehra.

She moves on upstairs. The doorbell rings.

[*As she disappears*] Will you get that, Tony?

TONY *comes out of the kitchen, sees* ZEHRA *standing immobile with tears in her eyes, frowns and goes to the door.* CONNIE *enters. She moves towards* ZEHRA.

CONNIE: [*to* ZEHRA] Sorry I couldn't make it sooner. I had two appointments that I couldn't shift.

ZEHRA *doesn't seem to hear her.* CONNIE *looks at* TONY.

What's going on?

TONY: She's terrified. She worries all the time that she's said something wrong to Mrs Blasko.

CONNIE *moves across to* ZEHRA *and sits down next to her.* TONY *goes outside.*

CONNIE: Zehra, what's happening here?

ZEHRA: I told Mrs Blasko about the scarring.

CONNIE: What scarring?

ZEHRA: [*tearfully*] Sarila fell off the bunk and gashed her cheek very badly. The scar could be there forever.

CONNIE: Oh, God. I'm sorry.

ZEHRA: Mrs Blasko is under stress. She doesn't want to be bothered with my problems.

CONNIE: No, Mrs Blasko's focus is very firmly on Mrs Blasko.

ZEHRA *breaks down in distress.* CONNIE *comforts her.* CARMELA *hears the noise and reappears on the stairs.*

CARMELA: What's going on?

CONNIE: Zehra is terrified out of her mind because she 'bothered' you about Sarila's scarring.

CARMELA: For God's sake, she's totally overreacting. If there's something seriously wrong with her children, of course I want to hear. Why are you here?

CONNIE: Because Zehra called and said my father's locked himself in his room.

CARMELA: Then stop pressuring him to confess. Get him institutionalised. And if you don't mind, leave me to deal with my own staff.

CONNIE: [*indicating* ZEHRA] She's a mess.

CARMELA: It's none of my doing.

CONNIE: Are you kidding?

CARMELA: [*in a tantrum*] She's in a bloody trance most of the time. What am I expected to do?

CONNIE: Maybe talk to her?

ZEHRA: Please. No. Please. It's fine.

CONNIE: Zehra, it's anything but fine. You're terrified of everything you say.

ZEHRA: Please. Connie, Mrs Blasko. Things will be fine now.

CARMELA: They better be. [*To* CONNIE] Now will you go and tell your father that my patience has finally evaporated. He's going.

> CARMELA *disappears back upstairs.* ZEHRA, *distressed, moves out into the kitchen.* CONNIE *moves to follow her but* ZEHRA *shakes her head violently. The front door opens and we hear* VIVIENNE *talking breathlessly to* ZIGGI.

VIVIENNE: [*offstage*] My Economics teacher Mr Evans is a total dud. I'm not kidding, but like, and this is true—I already know more about futures dealing than he does. I ask him three fairly simple questions and it's—Duh?

> *They both appear.* VIVIENNE *is so deeply engrossed in her stream of consciousness she doesn't see* CONNIE *at first, but* ZIGGI *does.*

I mean, how crazy is that? This schmuck is being paid, what, sixty thousand p.a.?— Oh, and Legal Studies— That idiot Mr Longworth got it totally wrong about the Restrictive Trade Practices Act. Totally wrong. Oh, hi.

VIVIENNE *bounds across to her aunt and kisses her on the cheek.*

ZIGGI: Connie. Leave Dad alone.

CONNIE: He's locked himself in his room.

ZIGGI: Not again.

CONNIE: Zehra says he's not eating.

ZIGGI: He's driving us crazy.

CONNIE: Use some of your fortune and put him in a high-quality village with back-up care.

ZIGGI: I'd send him to the Ritz bloody Carlton but he won't bloody go.

CONNIE: He's trying to starve himself to death.

VIVIENNE: [*at high speed*] He won't. He just feels totally trapped because he thinks Dad is going to go ballistic if he confesses and Auntie Connie's going to go ballistic if he doesn't. The hunger strike is just a signal to you both. In fact, having studied quite a bit of psychology myself lately, I'd say he's still suffering post-traumatic stress disorder.

CONNIE: Thank you, Vivienne. I had picked that up. You seem very buoyant.

VIVIENNE: I feel great. My only problem is that I'm showing real potential in so many areas that I can't even begin to decide what I really want to do. Hey, did you read that stuff today about that young Australian fashion designer that, like, just took Paris by storm? And by the way, fashion is another possibility for me. Always had a good eye for it. Anyrate, she said no one out here took any notice of her whatsoever. Hawked her designs round all the top houses and they told her her stuff was crap. But she believed in herself and off she went to Europe and three months later she's a superstar, and so this reporter goes back to all those geeks who think they're so shit-hot out here and says, 'Hey you turned this girl down', and, like, instead of saying, 'Yeah, we stuffed up', they all said that the guys in Paris had got it wrong. Like, that somehow Paris is past it and Australia is now the centre of fashion. Like, give me a break. How pathetic can you be? It made me think that I've got to get the hell out of this backwater as soon as possible and get across to New York or Europe, before I run up against the great Australian mediocrity. [*She bounds towards the stairs and up them, still talking.*] Just going to get on the net and see how my simulated portfolio is doing. Oh, and Dad, I'm working on a very, very special project that will make you very, very proud of me.

ZIGGI: What?

VIVIENNE: You'll see. You'll see.

> *She disappears into her room, still talking to herself.* ZIGGI *looks at* CONNIE *contemptuously.*

ZIGGI: That's the girl who's depressed, huh?

CONNIE: Ziggi, I hate to tell you this, but Vivienne's in an acute manic phase. She's bipolar.

ZIGGI: Not more of this shit!

CONNIE: She should be on Lithium.

ZIGGI: Hey! The girl is bounding with life, taking on the world and you want to turn her into a zombie? Get out of here.

CONNIE: She's in a state where her estimate of her abilities is about ten times more than it should be. It's odds on she'll do something totally reckless.

ZIGGI: Get out of here.

CONNIE: Ziggi, she could do anything.

ZIGGI: Like make a huge success of her life. That would be terrible, wouldn't it? Now would you kindly go.

CONNIE: I'm here to see Dad.

> MARKO *appears. He seems to have shaken off his depression. He stands upright, gimlet-eyed, and looks his offspring in the eye.*

MARKO: About what?

CONNIE: You've been locked in your room.

MARKO: I've made a decision. [*Calling loudly*] Vivienne! Vivienne!

> VIVIENNE *appears on the stairs.*

VIVIENNE: Yes, Grandpa.

MARKO: Is it coming?

VIVIENNE: [*nodding*] I ordered it on the net.

MARKO: Exactly as I specified.

VIVIENNE: They got an expert on it. It'll be perfect.

MARKO: Better be for the money I paid.

VIVIENNE: And Grandpa. I respect what you're going to do.

> VIVIENNE, *still hypermanic, disappears up the stairs again.*

ZIGGI: Dad, what the hell have you ordered?

MARKO: You'll find out.

ZIGGI: If you do anything stupid it'll finish my career.

MARKO: Then maybe you might do some real work. Like laying a concrete driveway on a twenty-five degree upward incline on rocky soil when it's forty degrees in the shade. Not once. About five thousand times. Then you'd get some idea of what I did for you. And now you want me out of this house because I'm an embarrassment.

ZIGGI: Dad, just calm down.

MARKO: I'll go into that old people's home. It couldn't be worse than here.

> MARKO *goes back towards his bedroom.*

ZIGGI: What the hell's he going to do now?

CONNIE: I have no idea.

◆   ◆   ◆   ◆   ◆

SCENE THIRTEEN

Ziggi's living room. Some days later.

CARMELA *comes down the stairs, dressed to go out.*

CARMELA: It'll be just a few more minutes before the babysitter arrives, Zehra. I think I can hear Freda.

> ZEHRA *nods and goes.* CARMELA *moves across to the phone, settles herself into a comfortable chair, hesitates, then nervously picks up the phone and dials, anticipation on her face until someone answers.*

> [*Trying to muster as much confidence as possible*] William, I just rang to thank you so much for coming in to take a look at me the

other day. I was so thrilled that you were so encouraging and supportive. [*A pause as she nods her head, smiling.*] I hadn't heard from you, so I just thought I'd ring to talk about next season and how thrilled I'd be if you wanted me to be part of it. From what I hear it's going to be very, very exciting. [*She listens to William and the smile gradually fades from her face.*] William, you said my work was fantastic. Or did I mishear? [*She listens and gets more and more furious. She tries to control herself.*] William, I can't quite believe I'm hearing this. You might find one role for me in the season. And a minor one at that. [*Pause.*] I'm supposed to feel grateful? Can I remind you how grateful you were when I stepped in and saved the day in London? [*She listens a little longer with growing fury.*] You obviously think I'm past it. Mother with child, forget her. Well, can I just say that some of the talent you've been pushing is just not up to it? [*Beat.*] No, it isn't just my opinion, William. It's what everybody is saying and if you haven't been hearing it, then I'm sorry. You soon will. [*She slams down the phone, furious.*] Zehra!

ZEHRA *appears holding Freda.*

I can't eat out. I can't face it. Just get something together, will you? It doesn't have to be grand, but on the other hand I want it to be edible. Understand.

ZEHRA *stands there staring at her.*

Not this again. Zehra, did you hear me? Do you understand?

ZEHRA *stands there immobile.*

Zehra, if there's any more of this Easter Island statue stuff you'll just have to go.

ZEHRA: [*quietly, but with intensity*] Tonight I have to go home.

CARMELA: Excuse me.

ZEHRA: This was the first day back at school for Sarila since her accident. She was crying and crying. She didn't want anyone to see her cheek.

CARMELA: You'll be an hour later at the most.

ZEHRA: I will show you what food we have and you can cook it.

CARMELA *stares at her.*

Sarila was very, very upset.

CARMELA: She'll be fine. All of us collect little nicks and scratches.

ZEHRA: This is a big gash, right across there. [*She illustrates on her own face, the size of the gash.*] She thinks she's lost her beauty.

CARMELA: Zehra, I'm sure your little girl—what's her name again?

ZEHRA: [*suddenly quietly angry and determined*] Sarila. She's six and brave as a lion. A gash almost through her cheek. Blood everywhere. And she's calming me down, not the other way. And my two others are Cari and Yasemin. Cari is twelve. She's amazing for her age. She does the cooking. Organises the shopping. Never complains. Without her we couldn't carry on. Yasemin is ten. Now Yasemin does complain. She wants this, she wants that. She wants what all the other girls have got. She makes me sad but she also makes me smile. She has decided she is going to be very, very rich one day and who can blame her? They are my life. Now I am going home to see Sarila. Do you want me to show you the food? If you want me not to come back tomorrow, then so be it.

CARMELA: [*suddenly bursting into tears*] Zehra, you think I lead a charmed life. You think I have everything. I've just been told I'm finished. That all this relentless grind has been for absolutely nothing. It was only a handful of years ago that *The Times* in London said I brought a vitality and freshness to the stage that was like a breath of fresh air and now it's as if none of that ever happened. You can believe it or not, Zehra, but I was good.

ZEHRA: I can see from the photo.

CARMELA: I'm sorry about your children. I should have asked about them. Bring their photos. Please. Tomorrow bring their photos. This obsession had me totally in its grip. I was desperate to be that photo in the hallway again. Otherwise I had nothing.

ZEHRA: No one with a child like Freda has nothing.

CARMELA: Tell little… Carilla was it?

ZEHRA: Sarila.

CARMELA: That wounds inside are always much deeper than those outside.

> CARMELA *sobs.* ZEHRA *looks at her with a mixture of compassion and contempt. She moves across, puts Freda in* CARMELA*'s arms, picks up her bag and leaves.* ZIGGI *comes slowly down the stairs. He looks stricken.*

William's just ditched me. Offered me one minor role in the whole of next season.

ZIGGI: What did you expect? You can't stay at the top forever.

CARMELA: Ziggi. Couldn't you be a little understanding?

ZIGGI: Honey—if we're talking disasters—I've just had one that blows yours right out of the H-two-O.

> VIVIENNE *appears at the head of the stairs looking depressed and chastened.*

[*To* VIVIENNE] Tell her!

VIVIENNE: I wanted you to be proud of me.

ZIGGI: I just went to check my cash management account on the computer, and guess what? All gone. Vanished. One and a half million dollars.

VIVIENNE: They said they were only taking a hundred thousand.

ZIGGI: Unbelievable. She photocopied my signature and faxed an authorisation to withdraw *and* my access code to a shonky share dealer in Thailand!

CARMELA: Why did you give her your access code?

ZIGGI: I change it every week so I have to have it on record. And she found it, didn't she?

VIVIENNE: They said they were only taking a hundred thousand.

ZIGGI: Little Miss Genius here plugged into this fake website and bulletin board about this amazingly undervalued stock—and they suckered her in with forty-three percent return on her first little flutter and then asked for authorisation and access code.

VIVIENNE: I checked them out thoroughly.

ZIGGI: By emailing a list of satisfied customers that *they* supplied? Jesus H Christ! Are you totally stupid?

VIVIENNE: I thought I was making money for you. I wanted you to be proud.

CARMELA: Surely the police can do something?

VIVIENNE: Yes, they could put my daughter away for years for forging my signature.

CARMELA: The Thai police.

ZIGGI: The Thai police? They're probably running the scam.

VIVIENNE: I was trying to make you proud of me.

CARMELA: For God's sake, send her back to her mother's.

ZIGGI: Her mother won't have her! Which I now realise was a bloody good call. There's a boarder's place coming up at her school and when it does she's right out of here.

VIVIENNE: I was just trying to make your life less stressful.

ZIGGI: Thank you, Vivienne. The stress of managing a million and a half dollars has totally disappeared.

> VIVIENNE *retreats tearfully back up the stairs.*

You just better pray they sign my new contract or we're down the drain.

CARMELA: We've still got assets, surely?

ZIGGI: What assets? I was forced to give my last house to my last wife and we owe the bank a fortune on this place.

CARMELA: You told me that you only have to put in a good word for those corporate types on air and they shower you with cash.

ZIGGI: Thanks to a totally hysterical press even that's in danger.

CARMELA: You still earn a fortune.

ZIGGI: Honey, we spend a fortune! No, let's rephrase that. *You* spend. I'll show you one month's restaurant bills if you're interested. And what we spent on last year's trip to every known luxury chateau in France. And what you spend on your bloody wardrobe. Most of which you wear once then move on. When Imelda Marcos wants to borrow shoes she comes to you.

CARMELA: They're going to sign your new contract, surely.

ZIGGI: There's nothing sure about it. My ratings are still dipping and they're telling me that unless I go in hard and kick more arse I could be finished.

CARMELA: Connie said that she should've been on Lithium.

ZIGGI: Just shut up.

CARMELA: Your stupid daughter's lost you a fortune, but I've lost a life.

ZIGGI: Honey, for one brief instant, because two much better dancers were injured, you were top of the pile. That was your moment. Get real.

CARMELA: You are a vile, insensitive, right-wing lout, and I was crazy to ever marry you.

> *She gets up and runs up the stairs, carrying Freda.*

ZIGGI: Better have a few more lunches with Graeme Bastoni! Maybe he'll still have you and you can swan off to the opera and ballet every second night!

◆　◆　◆　◆　◆

SCENE FOURTEEN

Ziggi's living room. That day.

ZEHRA *is cleaning.* CARMELA *comes down the stairs.*

CARMELA: Don't forget the silverware.

ZEHRA: I brought the photos of my children.

CARMELA *stares at her.*

Like you asked.

CARMELA: Oh. Good. I'll have a look at them later. I'm just going out to lunch with a friend of mine. Graeme Bastoni. You've heard of him?

ZEHRA: No.

CARMELA: One of the richest men in the country. Oh, er, how is young Carilla's cheek?

ZEHRA: She's still upset. But she will learn to cope.

CARMELA: I'm sure she will. Is my hair all right at the back?

ZEHRA: You didn't really want to see the photos, did you?

CARMELA *frowns.*

CARMELA: Zehra, this is becoming impossible. I did want to see the photos, but not right now.

ZEHRA: I can't work here anymore.

There's a pause.

CARMELA: Zehra, frankly it'll be better for both of us. I know things are tight for you so I'll write you a cheque for a thousand dollars to tide you over until you get another job.

ZEHRA: I can't accept.

CARMELA: Zehra, I know I have my faults. I'm not gushing and oozy and I can't fake that like some people, but I do know that life is difficult for you, so please let me do this one thing.

She pulls out her cheque book.

Please, you're a good person. You've been terrific to Freda and I can't do the big hugs thing, so at least let me give you *money*.

ZEHRA *nods and takes the cheque.*

ZEHRA: The truth is I can't afford to say no.

CARMELA: You'll stay on until I can find a replacement?

ZEHRA: Yes.

CARMELA: I know there's absolutely no justification for me feeling so wretched most of the time, you're the one with the miserable life.

ZEHRA: My life isn't miserable. It's just hard. But I have three daughters that are my joy.

CARMELA: My daughter's a joy too, so why's my life so miserable?

ZEHRA: [*angry*] Because you are married to a bully who has nothing but hate in his heart. And you stay with him because you want to live like this. And if I've hurt you, I'm sorry, but finally I don't care anymore.

CARMELA: You really don't like me, do you?

ZEHRA: Take care of Freda.

The two women look at each other. CARMELA *turns away and proceeds to the kitchen.*

CARMELA: [*offstage*] Tony, where are you? I'm late.

ZEHRA *throws away her mop and sits on the nearest chair.*

◆   ◆   ◆   ◆   ◆

SCENE FIFTEEN

The radio studio and Ziggi's house. Some weeks later.

ZIGGI *waits for his 'easy listening' selection to finish, tapping his fingers even more impatiently than normal.*

At Ziggi's place, MARKO *listens to his son.*

ZIGGI: That was the great Wayne Newton, and this is Ziggi Blasko on Life 608, the station that lets you get it off your chest. One thing that really gets up my nose is how our whole state education system is ramming guilt-unlimited down the throats of our young kids. The black-armband nation gone beserk. Did we steal this land away from its original inhabitants? Yes. Did the Angles, Saxons Jutes and Danes steal England from the Celts? Yes. And did the Normans steal it from them? Yes. And was just about every country in the world overrun and stolen in the course of its history? Yes. Welcome to world history. Were Aboriginals massacred? Undoubtedly. From time to time some farmers who'd lost one of their family members or dozens of their sheep to an indigenous spear, went on a rampage of retribution. But not that many and not that often. Let's all get over this national breast-beating guilt trip and get on with life. Ron from Blacktown.

RON: [*voice-over*] Ziggi, has anyone ever heard of a country called Germany? Just over fifty years ago they slaughtered millions of their own citizens. And do Germans spend their time breast-beating? No. Not if they had nothing to do with it.

ZIGGI: Precisely, Ron.

RON: [*voice-over*] If there are any Germans or their fellow travellers still living, who can be shown to have actually participated in Nazi slaughter, then they deserve everything coming to them.

ZIGGI: Yeah, but they'd be so old—is it really worth it?

RON: [*voice-over*] Old or not, they should pay. I'm not arguing leniency for real war criminals, but I am saying stop blaming their descendants.

ZIGGI: You've got a point there, Ron. Let's have a bit more music.

◆   ◆   ◆   ◆   ◆

SCENE SIXTEEN

Ziggi's place. Later.

MARKO *is putting on a German uniform.* ZIGGI *comes in the front door and down the passageway and stares at him.*

ZIGGI: Where in the hell did you get that?

MARKO: I ordered it on the internet.

ZIGGI: What in the hell do you think you're doing?

MARKO: The press are coming.

ZIGGI: Press?

MARKO: To hear my confession.

ZIGGI: Wearing that?

MARKO: I want to get maximum coverage.

ZIGGI: In my house? Photos of you in a Nazi uniform in my house?

> *The doorbell starts to ring repeatedly.*

MARKO: Here they are. Let them in.

ZIGGI: Dad, this could be the finish for me… Can't you understand?

MARKO: I'm sorry. The world must finally know what I've done.

> MARKO *moves towards the door.* ZIGGI *moves rapidly towards the back of the house as voices are heard in the hallway.*

◆　◆　◆　◆　◆

SCENE SEVENTEEN

Ziggi's living room. Some weeks later.

CONNIE *enters with Marko's case. She prepares to leave with* MARKO. VIVIENNE *has her case packed too.* ZIGGI *snaps at them belligerently.*

ZIGGI: If your aim was to finish me off, you failed.

CONNIE: Finishing you off was not the exercise.

ZIGGI: It all backfired on you, didn't it? He didn't know their names, he couldn't even remember the place it was supposed to have happened, none of his cohorts were still alive, and there were no bloody witnesses. They just thought he was a lunatic.

CONNIE: You told them he was a lunatic.

VIVIENNE: At least he faced up to what he did. He tried.

ZIGGI: I don't want to hear any moralising from you after what you did.

VIVIENNE: Dad, I'm sorry for what happened but it wasn't me who did it. Not the me who's standing in front of you now. Now I'm on medication I'm responsible for my actions again.

ZIGGI: Great, but it's a tad too late to be of any use to me!

VIVIENNE: If you're going to keep holding it over me for the rest of my life then where do we go from here?

ZIGGI: [*to* VIVIENNE] I'll tell you where we go. You get nothing from me from here on in. Not a cent!

VIVIENNE: Good, because I wouldn't take it.

She leaves in self-righteous high dudgeon. MARKO *appears.*

ZIGGI: [*to* MARKO] Lucky you didn't shoot Jews. Mossad would've had you in a dungeon in Tel Aviv weeks ago. I hope you feel better, Dad. I hope you do.

MARKO: I do.

ZIGGI: All that happened was that my name was plastered all over the media. Well done.

MARKO: I did what I felt I had to do.

ZIGGI: [*to* MARKO] Go back to Connie. Good riddance to the lot of you. I don't care if I never see you again.

CONNIE: Frankly, neither do we.

They go. ZIGGI *stands there fuming, then collapses in a chair.*

◆　　◆　　◆　　◆　　◆

SCENE EIGHTEEN

Radio studio. A week later.

ZIGGI *waits for his 'easy listening' selection to finish, tapping his fingers more impatiently than normal.*

ZIGGI: That was Harry Connick Junior, one of the few vocalists around that still knows how to sing in tune. And this is Ziggi Blasko on Life 608, the station that lets you get it off your chest. I sat down and thought about my life the other day, and suddenly I felt disgusted with myself. I've let myself get too soft, too mushy, too careful. The dreaded fog of political correctness which envelops and chokes us all was wearing me down too. Well, sorry, as from today you're getting the old Ziggi Blasco back. The one you could trust to tell it like it is. One of my really pet hates is phoney feminism and its breathtaking double standards. Get this. The feminists are saying that men have no right to check the paternity of their children. If the missus walks out on her guy and he suspects the maintenance money he's paying her is going to a kid who isn't his, he's not allowed to check it out with DNA. He's got to take her word. Get that, her word. Why? Because if she did screw around behind his back with her personal trainer or the pool cleaner or whoever—it's his fault. Can you believe that? His fault. Why? Because he didn't fulfil her. Physically and emotionally. He failed her, so it's all his fault. Our partners can ignore their wedding vows and do what they like behind our backs and we pay for the resulting children all our lives. What kind of justice is that? I'll tell you. It's not justice, it's a travesty of justice. And it's just one instance of how women, gays and any other group that can shriek loud enough are remaking the world to suit themselves, and to hell with the decent hardworking guys whose only sin is that they get on with their lives without complaining. I'm outraged. I'm livid. Let's hear what you've got to say. Virginia from Birchgrove.

VIRGINIA: [*voice-over*] Ziggi, are you saying that every child should be tested? That no women can be trusted?

ZIGGI: Virginia. My wife can be trusted. A lot of wives can be trusted. But any man who has good reason to believe that his wife can't be trusted has a total right to check that out. And you can argue otherwise till the cows come home but, Virginia, you're wrong. Just plain wrong.

◆   ◆   ◆   ◆   ◆

SCENE NINETEEN

Ziggi's living room. Later that evening.

CARMELA *sits in the living room holding Freda. The front door opens.* ZIGGI *comes bounding in, elated.*

ZIGGI: Hey, did you hear me today?
CARMELA: No.

ZIGGI: I bored it up the feminists. I was on song. Totally on song. The response was enormous. The Chief called me in and said my contract will be renewed with a substantial increase. We're fine. We're going to survive.

CARMELA: Ziggi. I'm moving out.

ZIGGI: You're what?

CARMELA: I can't stand it anymore.

ZIGGI: Can't stand what?

CARMELA: You, our life. Everything. Tony!

TONY: [*entering*] Yes, Mrs Blasko.

CARMELA: Are the cases in the boot?

TONY: Everything's loaded.

ZIGGI: This is incredible. On air today I said I was lucky to have a wife I can trust and a few hours later you tell me you're walking out! Where the hell do you think you're going?

CARMELA: To my sister's. We'll talk about this later.

ZIGGI: Hey, hell no. We'll talk about this now! You can't just walk out and take my child.

CARMELA: She mightn't even be yours!

ZIGGI: What!

CARMELA: It's not my fault. You rushed me into marriage before the thing with Graeme was ever really resolved.

She starts to walk out. ZIGGI *stands there dumbstruck.*

Tony, bring my things. From now on you work for me.

TONY: You know what? If I was heading up Al Qaeda I'd set a camera up right in this house. Half an hour of footage of you two and the suicide bombers would be begging to be sent here. 'Forget the Jews, and the Yanks', they'd be howling. 'Get over here, guys.' 'This is the country where the real pricks hang out.' 'You think Western culture has anything to offer? Check out this video.' [*To* CARMELA *and* ZIGGI] You two don't give a shit about anything or anyone except your own giant egos, and the sound of your own voices. And you pay rock-bottom wages and abuse your servants for taking home a few scraps of food. Sorry, Mrs Blasco. As from this moment I don't work for either of you.

He storms out. CARMELA *looks shocked but follows.* ZIGGI *is dumbstruck.*

◆　◆　◆　◆　◆

SCENE TWENTY

Radio studio. Next day.

ZIGGI *waits for his 'easy listening' selection to finish.*

ZIGGI: That was the great smoothie Tony Bennett, and this is Ziggi Blasko on Life 608, the station that lets you get it off your chest. Well, I've been touched, even overwhelmed by your messages of sympathy since the gutter press decided to put the knife into me. The tall poppy syndrome is still alive and well out there and I seem to be the number one target. Yes, my daughter did nearly bankrupt me. Yes, my wife did leave me for Graeme Bastoni, a much richer man than I'll ever be, taking my child with her. And, yes, my father did his level best to totally discredit me. But you know what? That which doesn't kill you makes you stronger. And no matter how black things are there's always someone who redeems your faith in human nature. I got a letter from a lovely lady we employed as a housekeeper and nanny called Zehra. She enclosed a cheque for one hundred and fifteen dollars which she says she owed me because she sometimes bought more food than we really needed and took some home to her own children. That's what I call honesty and decency and you don't see it too often these days. I got my producer to hunt up Zehra and she's on the line now. Zehra, are you there?

ZEHRA: [*voice-over*] Yes I'm here, Mr Blasko.

ZIGGI: Thank you for the cheque. I was really impressed with your honesty.

ZEHRA: [*voice-over*] I'm sorry. It took me a while to save it, but it has been worrying me a lot.

ZIGGI: It needn't have. You did a wonderful job. But I do appreciate your honesty and I'm really pleased to think that there still is some decency out there.

ZEHRA: [*voice-over*] I didn't want you to think Muslims are dishonest.

ZIGGI: There are good Muslims and bad Muslims. Unfortunately it does seem that the bad ones try their hardest to win out. Zehra, we paid you the award rate, but things were still tight for you, right?

ZEHRA: [*voice-over*] Right.

ZIGGI: You could be categorised as what our bleeding hearts call 'the working poor'. But your children were well-fed, happy, doing well at school and enjoying what this society has to offer, right?

ZEHRA: [*voice-over*] My children were happy. Mostly. I could never afford school excursions, but they understood.

ZIGGI: But they were well-fed?

> ZEHRA *appears on stage. She's scared. She's literally shaking with fear. She nods as if about to agree, but something within her rebels.*

ZEHRA: No. Not so well-fed, but I'm not here to complain.

ZIGGI: Well, I mean not steak every night, but well-fed.

ZEHRA: No, never steak. When I was working out my spending, food came last.

ZIGGI: Zehra, come on. That can't be right. Food came last?

ZEHRA: Sorry. That makes it sound worse that it was. My children had lots of toast and coffee in the morning, so they could get by without lunch. And four nights a week we had just vegetables and rice or noodles and soy. But I made it taste good. And on three nights we did have mince or tuna, which was exciting for all of us. And once a week fruit.

ZIGGI: Yes, but no lunch? For heaven's sake? What was your rent?

ZEHRA: Two hundred and forty dollars.

ZIGGI: Surely you could get cheaper than that further out?

ZEHRA: Mr Blasco, we were about as far out as we could go. It took me almost two hours by train and bus. I had to get up at five and didn't get home till nine. You know that. But, please, I didn't want to say these things.

ZIGGI: Well, you have, Zehra, you have. You're turning what was a really decent gesture into one great whinge.

ZEHRA: I was just answering your questions.

ZIGGI: No, Zehra, you were making accusations.

ZEHRA: [*suddenly emotional and angry herself*] You want accusations? Fine. Working for you was terrible. Terrible. I began to understand the hatred that led women to strap explosives around their waist. I only stayed as long as I did out of fear of being unemployed, because for me, even one week out of work is a catastrophe.

ZIGGI: Sorry, Zehra, but you're off the air. And I'm pretty disgusted, I must say, that you turned what started as a decent impulse into a giant whining session. Let's hear John Williamson's great Australian anthem *True Blue*.

ZEHRA: [*continuing to talk over the song*] I am working for a family that pays me a little better. I don't work crazy hours. My children have meat at least four nights a week, and I am slowly saving for a holiday. Not a grand holiday, but it will be our first ever and we talk about it every night. Ziggi, I have paid my final debt to you. I hope I never ever hear your voice again.

> ZEHRA *stands contemplating what she just did. She's still shaking with fear but she also shakes her head in affirmation. She did what had to be done and she's glad.*
>
> *On another part of the stage* CONNIE *and* MARKO *have been listening to the exchange.*

CONNIE: [*gleeful*] He's on the way out, Dad. He's finally on the way out.

MARKO: No.

CONNIE: Dad, he's a dinosaur. He's on borrowed time.

MARKO: Connie, you are intelligent, but Ziggi is smart.

CONNIE: Rat cunning.

MARKO: No, smart. You think that if you teach tolerance it will happen.

CONNIE: We can change, Dad. We are.

MARKO: I grew up with Serbs as my neighbours. My intelligence told me they were as human as I was. But Hitler was smart. He knew the buttons to press. Ziggi knows the buttons to press.

> *In the radio studio,* ZIGGI *waits, still angry.*

ZIGGI: That was the wonderful Gypsy Kings and that's just about it for today. Well, that was interesting. Our ex-housekeeper. Supposedly a moderate Muslim, but scratch the surface a little and suddenly she can 'understand' why Muslim women become suicide bombers. It doesn't take much for the true colours to be revealed. Australia and the Western world, take note and be warned. Don't go away. Ziggi

Blasco will be back. Ziggi Blasco will always be back, because this country wants to hear what I've got to say.

The show's end music plays. ZIGGI *stares out into the audience, confident that he's right.*

THE END

OPERATOR

Henri Szeps (left) as Douglas and Rory Williamson as Jake in the 2005 Ensemble Theatre production. (Photo: Steve Lunam)

The Operator

Colleen Ryan

The scale of the betrayal that David Williamson presents in this tale of corporate bastardry is breathtaking. But it is only so because of Williamson's own clever use of scale. Setting *Operator* in such a small company grants a certain clarity to the moral issues at stake. In a larger corporation, with plenty of bodies in between to soften the edges of the deceit, this behaviour is quite normal.

Corporate Australia is littered with Jakes—many of them quite well-known Chief Executives and Chairmen. Others are household names largely due to their spectacular fall from monied grace.

There are touches of Machiavelli's *The Prince* and Sun-Tzu's *The Art of War* in this unpalatable portrait of the workplace. But Williamson has placed his exposition in a much more contemporary, and ordinary, setting—a manufacturer of gym equipment for 'travelling and execs and career women dreaming of the buffed, sleek bodies they're never going to inhabit'.

Williamson, in unmasking the blight of corporations, has given a succinct blueprint of what is required to be an 'Operator'. Absolute amorality; arrogance; a mixture of obsequiousness and bastardy which can be turned on and off at will; perfect timing; and the rat cunning to trade on the weaknesses, and the decency, of others.

In a depressingly familiar pattern, the Operator, Jake, moves on to a bigger and better role leaving financial disaster and destroyed careers in his wake. We have seen this happen again and again in Australian companies. Recruitment consultants and Boards of Directors don't seem to get any smarter when it comes to choosing CEOs. Eventually, they get found out. Sometimes they die prematurely and the mess in the bottom drawer is discovered. For example, Robert Holmes a Court at the Bell Group. Sometimes the regulator catches up with them, as they did with Laurie Connell at Rothwells merchant bank. And sometimes they simply reach their true level of incompetence and the market works—it slays them. It slays them—via the share price, the sharp tongues of analysts and hopefully the financial press. There are dozens of examples, but unfortunately due to Australia's strict defamation laws they can't be detailed here.

Straight out fraud is often part of the equation—although it is rarely considered fraud until the perpetrator gets caught. Instead it is seen as sharp business practice. Sailing a bit close to the wind. These are mere euphemisms for clear theft or misrepresentation with financial consequences.

Williamson has very, very cleverly woven together the characteristics of a Grade-A corporate bastard and placed him in a small office environment. In the process he has also touched on the wider issues of nepotism, sexual ambiguity, youth versus age, the dangers of email, the threat of China, the utter heartlessness of an international corporate takeover, the importance of relationships over work.

There are subtleties too in Williamson's treatment. Nepotism is unfair—but it creates its own victims, like Irena as the niece of the Chief Executive, Douglas Scrivener. Irena represents those who are given a 'leg up' without their own participation, perhaps against their will, and then are left to suffer the derisory comments of their colleagues.

The plague of sexual differences is such amidst the conservatism of the corporate world that it can lead to dangerous sensitivity to the views of others. Once again, it is Irena, a lesbian, who demonstrates this in Williamson's treatment. And, of course, it is Jake, the 'Operator', who makes the scathing comment: 'Why doesn't she face up to reality? Her girlfriend left her for someone slightly less ugly than she is.'

The hubris of many who have succeeded and their vulnerability to like-minded souls is well represented here in the character of Douglas Scrivener: '… as long as I'm sitting in this chair, my call is your reality', a boorish Douglas tells Alex. The portrayal of Douglas presents the audience with a version of the truism—'the easiest man to con is a conman himself'. As the unsuspecting Douglas tells Jake, 'The manager who surrounds himself with ingratiators and arselickers is doomed'.

And then there is the protection that has been lost through the spread of electronic communication. But as Williamson cleverly details—when deception is the driver, old-fashioned methods, such as simulated signatures, can work just as well.

But what about corporate loyalty? Jake has something to teach Douglas here. '[This company is] small beer, Douglas. If you'd had any sense you would've got out years ago. You're never going to make the *Fortune* Top Two Hundred hanging around this dump.'

This play is a gem—small, perfect and multi-faceted. And it goes right to the heart of the dark side of corporate behaviour.

Shanghai
March 2005

Colleen Ryan is the China correspondent for the *Australian Financial Review* based in Shanghai. She was the editor of the *Australian Financial Review* between 1998 and 2001, and is the co-author of *Corporate Cannibals: The Taking of Fairfax*. She has won several awards for investigative business journalism including the Shareholders Association Award; the Graham Perkin Award for Australian Journalist of the Year in 1992, and two Walkley Awards.

Playwright's Note

David Williamson

Usually when I write a play it is because of something I have observed around me. For instance, in the case of *The Removalists*, one of my early plays, I was shifting homes and the removalist told me about this great day that he had last Friday and I thought, 'Oh...'. With *Travelling North* I borrowed heavily from the life of my mother-in-law and her partner whom I met on the coast of NSW in the early 1970s. It wasn't until about 1979 that it came into focus as a situation that would be good to write about, however. So, experiences don't always turn into plays immediately.

This is the case with *Operator*. Some years back I read an article, I forget where now. It concerned studies of corporate executives who had been given a battery of psychological tests to determine what kind of people they were. One of these tests was politely called the N Mach scale which is actually the Machiavellian scale. In simple terms, it measures one's propensity towards being a sociopath. For those of you who don't know, a sociopath is someone who has very little fear or anxiety, and no capacity for empathy.

They normally comprise about three or four per cent of the male population and one per cent of the female population. They are typically fearless and charming, with absolutely no anxiety in social situations. So they might come up to you in a foyer and start a charming conversation about something. The other side of the coin is that they are entirely manipulative. They view other people as objects to be manoeuvred for their own material ends. They have no remorse, no capacity for empathy, no compassion for any harm they cause to other human beings in their relentless drive for power and material success. It really is relentless, and only material success counts. Other sorts of success, such as having good and loving friends, a good and loving family, aren't on their radar. They are rather chilling human beings and we've all met some in life. With three in a hundred, you are bound to meet them.

The results of the N Mach scales for top-level executives surprised me. There were roughly eight times the number of high N Mach individuals in top levels of corporate life compared to the general population. In fact, around twenty-five to thirty per cent of our corporate executives exhibited sociopathic tendencies. And, the article went on to say, because they are so devious, such people are rarely discovered until it is too late. They present well, they are very confident, they are risk takers. They appeal to Boards who think, 'Aah, here's a go-getter. Here's someone who will be very good for our firm.' It often takes years to find out that they have been wrecking the firm from the inside. And it often doesn't happen until their unscrupulous risk taking and devious business practices create a spectacular failure. Of course over three quarters of top executives are decent and principled people who are just as bewildered about this type of behaviour as anyone else. But I found it chilling that sociopaths are so successful in our corporate culture.

So I decided to write a play in which the central character is a young, high N Mach male setting out on his corporate career. He starts at a rather modest firm and proceeds to feather his own nest and further his own interests. In the process he almost wrecks the company. This is typical. Despite this, though, his social flair and energetic, positive, risk-taking behaviour is spotted by some head-hunting firm who promote him to some other firm where he repeats the cycle. Again this happens all too often.

Recently we've had some of the biggest collapses in our corporate history. The collapse of firms such as Enron and HIH are phenomenal because their executives got away with financial murder. They appeared so confident. They could waltz into a boardroom and say, 'Everything's fine. You might have heard some rumours that there are a few problems but believe me we have it one hundred per cent under control.' So they weren't caught. And who suffers from such collapses? It's only occasionally the executives themselves. With a phalanx of highly-paid lawyers to protect them they seldom go to jail. Generally they've salted away enough money through bonus schemes and payments to come out in top shape. The ones that suffer are shareholders and employees.

The employees suffer in two ways. They've been subjected to Machiavellian behaviour and they're out of a job. So I wanted to take a character with charm, assurance, plausibility, and show how often in modern corporate life it is not real quality and performance that counts, it is the appearance of surety, flair and confidence. Business runs on confidence, so if you present confidently the chances are that you will be perceived as performing confidently even if you are not. So much of our corporate life is showbiz.

I remember going to Los Angeles when I was writing some screenplays there. I was told that what counts in America is not that you're a good writer, but that you seem to be a good writer. So you have to come into a room full of beady-eyed executives and sell them your qualities as a writer. Tell them how brilliant you are, how you are going to do the script. Then you might get the job. But if they sense a lack of confidence, right away you're out. I didn't last long in that environment because I'm not sociopathic enough.

That's the basis of the play. The rise and rise of Jake and the devastation he leaves behind him. Yet he doesn't suffer. I was trying to show in this play how justice is so often not done. To clobber the villains at the end of every play and say nobody ever gets away with it is just plain wrong. They get away with it time and time again.

April 2005

Operator was first produced by Ensemble Theatre Company at the Ensemble Theatre, Sydney, on 6 January 2005, with the following cast:

ALEX	Michael Ross
DOUGLAS	Henri Szeps
FRANCINE	Amanda Crompton
JAKE	Rory Williamson
MELISSA	Katrina Milosevic
IRENA	Melissa Gray

Director, Sandra Bates
Set and Costume Designer, Mark Thompson
Lighting Designer, Sydney Bouhaniche

CHARACTERS

ALEX CARMIDES

DOUGLAS SCRIVENER

FRANCINE

JAKE JENINO

MELISSA

IRENA

ACT ONE

ALEX: [*to the audience*] I'm Alex Carmides. Sometime head of product development in a medium-sized manufacturing company that makes inclined motorised treadmills, exercise bikes, rowing machines and steppers. The kind of stuff you see being used in hotel gyms by travelling execs and career women dreaming of the buffed, sleek bodies they're never going to inhabit. The products are fine, but there's only so many hours in your life you can step endlessly up hills that aren't there, and take long bike rides to nowhere without finally feeling really silly. But that wasn't my problem. I just had to make sure that our products improved at a faster rate than our competitors, so that our market share went steadily up. Which most of the time it did, because I worked at it. According to my ex-wife, I'm a workaholic emotional cripple. She left me quite a while back. I'm trying to repair things with my two sons who haven't seen all that much of me either.

I reported to Douglas Scrivener, our CEO, and the start of the chain of events that you're about to see, can be sheeted back to a day over a year ago when he called me to his office.

> ALEX *walks into an office where* DOUGLAS *sits behind a desk.* DOUGLAS *looks up and indicates a chair.* ALEX *sits, and waits.* DOUGLAS *surveys him. It's a technique of intimidation that he uses frequently. 'I'm the boss, sit still and wait.' Then* DOUGLAS *picks up a printout and hands it to* ALEX. ALEX *reads it. He grimaces and shrugs.*

DOUGLAS: Second month in a row we've lost market share.

ALEX: Very minor dip this month.

DOUGLAS: I hoped you wouldn't say that, Alex.

ALEX: Sorry?

DOUGLAS: I hoped you'd be as alarmed as I am. One month trending down could be chance factors. Two months means we're sliding. And you know something? That makes me feel anything but complacent.

ALEX: Douglas, I'm not complacent.

DOUGLAS: I'm alarmed. I assumed you would be too.

ALEX: Douglas, we're talking less than half of one percent.

DOUGLAS: Alex, the minute you relax in this game, the game is over.

ALEX: I'm not relaxed.

DOUGLAS: People who exercise are a fraternity. They exchange notes. They compare and contrast. You know what I think they're saying out there?

ALEX: No, I— [don't know what]

DOUGLAS: I think they're saying that our C-47 stepper hasn't got enough exercise programs.

ALEX: The stepper has got thirty different programs, and our— [research shows]

DOUGLAS: The American Step King Three has seventy-two.

ALEX: —and our research shows that ninety-eight percent of the users only ever use five or six.

DOUGLAS: Alex, there's one word that drives all human progress. Do you know what that word is?

ALEX: No.

DOUGLAS: Choice. People might never *exercise* that choice, but by God they want to know it's there.

ALEX: I proposed a sixty-program version a year ago, and I couldn't get an answer out of you.

DOUGLAS: It added twenty percent to the cost of the unit.

ALEX: It still would've retailed cheaper than the Step King Three.

> FRANCINE, *Douglas's PA, an attractive, efficient woman in her thirties, enters.*

FRANCINE: Sorry to interrupt, Douglas, but Frank and Michelle are here for the meeting.

DOUGLAS: I'll be a few minutes. Give them coffee. And one for me.

FRANCINE: You've had three already.

DOUGLAS: Francine.

FRANCINE: Did you read that article I left on your desk? Caffeine damages your health in seventy-three separate ways.

DOUGLAS: If I believed everything you've left on my desk I'd never drink water and rarely breathe. Am I allowed mineral water?

FRANCINE: The salt levels are disastrous, and the radiation levels are too high, but it's your funeral.

She goes.

DOUGLAS: There was a time when personal assistants used to assist. Alex, two of your three most crucial lieutenants resigned within a week of each other. Why?

ALEX: [*to the audience*] Mainly because they couldn't stand Irena, Douglas's favourite niece, who lost her last two jobs and who Douglas foisted on me. But I couldn't say that. Nepotism is a way of life around here. Douglas's second son is currently causing havoc in production and his daughter laid waste to marketing until she got bored and went to Europe. [*To* DOUGLAS] They got good offers elsewhere.

DOUGLAS: Human resources have narrowed the short list to five. Do you want to be part of the interviewing process?

ALEX: [*nodding*] They *will* be working for me.

ALEX *leaves the office. He's angry.*

[*To the audience*] If he'd had his way I wouldn't't've even been there to scrutinise my new staff. He was a total control freak. The only reason he listened to other people's opinions was to rip them to shreds.

ALEX *moves to sit next to* DOUGLAS *as they both face one of the short-listed candidates,* JAKE, *twenty-nine.* JAKE *is well-groomed and confident.*

DOUGLAS: Jake, you've got very impressive qualifications. Masters Degree in Industrial Engineering and a First-Class Honours MBA.

ALEX: Your references are solid, but not exactly glowing.

JAKE: Energy and enthusiasm aren't welcomed in some workplaces.

ALEX: [*to the audience*] I'll be honest. Right from the start I didn't like this guy. [*To* JAKE] Can you remember some times when things went very well for you in the past?

JAKE: Yeah. I stuck my neck out at Blackwell's and said that their impeller design was years out of date. My boss told me to get lost. I went straight to the CEO, pitched my idea, and my design turned out to be the industry leader. I'm the single main reason they're still in business today.

ALEX: That's a pretty big call, Jake.

JAKE: Maybe, but I honestly believe it's true.

DOUGLAS: Then there's nothing wrong with saying it. Can't stand false modesty.

ALEX: Did you have any qualms about going above your boss's head?

JAKE: No. He was wrong.

ALEX: How did you get on with him after that?

JAKE: [*shrugging*] As you'd expect, but when I know I'm right I back myself.

DOUGLAS: [*nodding*] A good employee has to have the guts to challenge his superiors if he's convinced they're wrong.

JAKE: I'm results-orientated, not ego-orientated. For me the outcomes for the firm come first every time.

From left: Michael Ross as Alex, Henri Szeps as Douglas and Rory Williamson as Jake in the 2005 Ensemble Theatre production. (Photo: Steve Lunam)

DOUGLAS: I wish a few more round here thought like that, Jake. There wouldn't be so much energy wasted in petty politics.

JAKE: I've got talents and abilities, and I use them. Full stop. Everything else is bullshit.

DOUGLAS: Absolutely.

ALEX: Can you remember a time when things went badly for you in the past?

JAKE: Had my share of disasters with women.

ALEX: What kind of disasters?

JAKE: The commitment thing. They want it, I don't. At twenty-nine I've got plenty of living to do.

ALEX: A wife and family? That's not living?

JAKE: Not the kind of living I'm ready for just yet.

ALEX: A workplace experience that went badly?

JAKE: I put them out of my mind and move on. You can't change the past, but— [you *can* change the future]

DOUGLAS: [*nodding*] You *can* change the future.

The interview will continue, but we jump to the end of it, just after JAKE *has left the room.* ALEX *gets up and paces around. Worried.*

He's top material.

ALEX: Wouldn't admit to a *single* mistake?

DOUGLAS: He doesn't obsess about them. He moves on.

ALEX *is patently not convinced.*

He's confident. He's got balls.

ALEX: I checked his referees and yes, he did design one impeller that was an improvement, but it didn't save the firm.

DOUGLAS: Okay, so he blows his trumpet a few decibels too loudly, but that's better than someone riddled with self doubt. I say grab him. And quickly.

ALEX: He worries me.

DOUGLAS: Lots of things worry you, Alex. Grab him.

And the two of them are now in their next interview with MELISSA, *twenty-seven, a cheerful but somewhat nervous young woman.*

Your qualifications are impressive, Melissa. Industrial Design, MBA.

MELISSA: My MBA got a little rocky at one stage, but I finally got there.

ALEX: [*to the audience*] There are some people I warm to immediately and Melissa was one of them. It wasn't anything sexual. To be honest, her dress sense was on the daggy side. She just had an appealing openness and vulnerability. That engaging thing of trying just a little too hard. The sort of person I would've liked as a daughter. [*To* MELISSA] You got Honours.

MELISSA: [*smiling*] Finally.

DOUGLAS: Why do you want to join us, Melissa?

MELISSA: Sometimes I watch SBS news and the horror of what's happening in the world really gets to me. I love the fact that your firm produces something that impacts positively on lives.

DOUGLAS: Could you elaborate?

MELISSA: There's mounting evidence that exercise is probably the single most effective way of increasing human happiness. It relieves stress and anxiety, improves sleep, slows the aging process, improves mental sharpness, reduces the risk of heart disease, relieves depression and enhances self image.

DOUGLAS: You exercise, Melissa?

MELISSA: No. I was studying so hard I never sort of found the time, which is no excuse really, I know, but if I get this job then I'll make sure I buy one of your products and get myself in shape. I really will.

ALEX: Can you remember some times when things went really well for you?

MELISSA: Oh gosh. That's put me on the spot. You know something, some of my best times were when I was really little. My parents were really loving and encouraged me to do things. I was a bit of a daddy's girl. Really spoilt I guess.

DOUGLAS: Something later in life?

MELISSA: You know it's a stupid thing, but when I made the netball team. Only the thirds, but it was really something because I wasn't great at sport. And I worked at it and made it. I remember them giving me one of those huge signs you put on your chest, GK, goalkeeper, and I looked at it and cried. I was good at blocking forwards.

ALEX: Can you remember some times when things went badly for you?

MELISSA: Try every second day. My love life, well, let's not go there.

DOUGLAS: No let's not. Some incident you recall?

MELISSA: One day in one of my MBA seminars. I hadn't done enough preparation and I was asked a question and made a mess of it in front of everyone and the lecturer said, 'Melissa, why did you enrol in this course? You're wasting your time and mine.' I tried to stop the tears but I couldn't and I was really down for weeks.

The interview will continue, but we jump to the end of it, just after MELISSA *has left the room.*

DOUGLAS: We can forget her.

ALEX: Hang on.

DOUGLAS: You *want* her working for you?

ALEX: She's got exactly the qualifications I'm looking for.

DOUGLAS: She apologises for herself. She gets 'down' for weeks. Alex, never hire a depressive.

ALEX: She was told she was wasting the lecturer's time and she ended up with an Honours Degree. She wasn't great at sport and she kept trying.

DOUGLAS: You want her?

ALEX: She's got better qualifications than the other candidates and... yeah, I think she'll be fine. I'll take Jake if you let me have her.

DOUGLAS: Be it on your own head then.

Some time later. ALEX *briefs his two new protégés,* JAKE *and* MELISSA.

ALEX: Okay, now I don't like to do this, but in this case I think it's necessary. The other member of the team you'll be working with is Irena, who is Douglas's niece. It's natural for relatives to talk to each other, so it might be prudent to assume that what you say in front of her might make its way back.

JAKE: What are her qualifications?

ALEX: A BA majoring in History.

JAKE and MELISSA *look at each other. Totally inappropriate qualifications.* JAKE *shrugs.*

MELISSA: Is she, I mean... er... nice?

ALEX: I think you'll have to make your own judgement on that one.

Some time later. JAKE *and* MELISSA *are discussing an upcoming meeting.*

MELISSA: Did you read that stuff I left on you desk?

JAKE: Yeah. Interesting.

MELISSA: Some of the new exercise products being developed in the States are really exciting?

JAKE: Yeah. Really ingenious.

MELISSA: I think we should be moving in that direction, but I'm not sure how frank I should be at the first meeting. I don't want to come across as a total know-all. What do you think?

JAKE: Say what you're feeling. That's what you're paid for.

MELISSA: It might seem as if I'm being very critical.

JAKE: Say what you're feeling.

MELISSA: [*still trying to talk herself into it*] I feel Alex's the sort of boss who'll listen.

JAKE: Maybe.

MELISSA: You don't think so.

JAKE: I'll wait and see.

MELISSA: There's something about him that makes me…

JAKE: What?

MELISSA: Feel I can trust him.

JAKE: [*raising his eyebrows*] Crush on the boss already?

MELISSA: [*blushing, embarrassed*] No!

JAKE: [*grinning*] Dangerous stuff.

MELISSA: [*still hugely embarrassed*] No. I just think he's got…

JAKE: A ten-inch wanger?

MELISSA: Jake!

JAKE: What's he got?

MELISSA: Integrity.

JAKE: Define it.

MELISSA: He'd say the same to your face as he would behind your back.

JAKE: How boring.

MELISSA: You're just horrible.

JAKE: Yeah, but incredibly magnetic.

MELISSA: Sorry, I just don't feel the pull.

JAKE: You're not my type either.

MELISSA: [*hurt, but putting on a brave face*] Good.

JAKE: Yes it is. We can work as colleagues, without the deep sexual vibes that are going to complicate things with you and Alex.

MELISSA: Stop it.

JAKE: Actually he does look a bit like that actor who does brooding magnetic unattainable males. What's his name again?

MELISSA: Jake, stop it.

JAKE: Colin Firth. Smouldering. Much more under the surface than it appears.

MELISSA: Jake, I've really, really had enough.

> *Some minutes later they're at the meeting with* ALEX. IRENA, *twenty-eight, is also there. She has a severe and unflattering hair style, wears clothes that make no concessions to conventional femininity and has a totally detached manner.*

ALEX: Jake, Melissa. Glad to have you on board. Apart from the everyday stuff which we'll talk about shortly, I'd like you all to meet at least once a week and identify possible areas of product change and fire your recommendations directly to me. I've got my own thoughts on where we should be headed but I'd like to be challenged and stimulated.

IRENA: Who'll chair this group?

ALEX: I'm quite happy for you to take on that role, Irena.

IRENA: No, no. They're the specialists. I just research and collate.

JAKE: I'm quite happy to do that if you guys would be happy with that.

MELISSA: Absolutely. I'm still feeling my way.

ALEX: That's fine then. Any questions?

MELISSA: When you say product change, does that include new products?

ALEX: I was thinking more in terms of improvements to our present product range, but if you've got new product ideas, sure. Did you have anything in mind?

MELISSA: I can't help feeling that our products are very conventional. Treadmill, bicycle, rowing machine, stepper. The old workhorses.

The research I've been doing indicates things are moving lightning fast, particularly in the US, and my fear is that we might get left behind.

ALEX: I guess our products could be called workhorses, but then again the word 'workhorse' implies they're reliable and used a lot.

MELISSA: Sorry, I didn't mean to sound dismissive. They're fine products and they'll always be the backbone of our range.

ALEX: What sort of new product were you talking about?

MELISSA: The Americans are coming up with a new exercise device just about every second week.

IRENA: Such as?

MELISSA: The Thigh 'N' Buns Buster, the…

ALEX: The what?

JAKE: A belt around your waist and two elastic strips that clip onto your ankles. Looks hideous.

MELISSA: [*thrown by* JAKE *pre-empting her*] It's effective. Your legs and buns have to work much harder.

IRENA: Where do you wear this contraption?

MELISSA: Everywhere. Shopping, housework, gardening—they're selling like crazy in the US. Then there's the Power Fitness Chute—you clip this little parachute on behind you when you run and it increases resistance.

JAKE: There's a version for swimming too.

MELISSA: There's the Body Blade. Really simple. Just this flexible big blade with two weights on each end and you hold it in your hands like this… [*showing them*] and move your hands up and down and the weights flap around and create resistance, and there's Flo, which is really simple—

JAKE: Sure is. A plastic bag with water in it.

MELISSA: [*jumping in*] As you move it up and down the water flows from one end to the other and provides a smooth counter resistance. There's the Pro Abdominator—

IRENA: Pro what?

MELISSA: [*fast off the mark so* JAKE *can't cut in*] These slings go from your shoulder to your elbow and hook onto an overhead bar and you can dangle and lift your legs up and down without any strain.

IRENA: Pro Abdominator? Sounds like an Arnold Schwarzenegger movie.

MELISSA: There's the Vassar Trainer which simulates swimming without ever having to get wet. And there's—

ALEX: Melissa. I appreciate the research, but frankly even if we *did* want to manufacture these exotic devices, they're patented.

MELISSA: No, I didn't mean we copy them. I just brought them up to show that it's possible to come up with all sorts of totally new exercise concepts.

IRENA: If I want to swim, I'll go to the local pool and swim.

MELISSA: You might as well say we shouldn't make steppers because you can find a mountain and climb it.

ALEX: Melissa, I really appreciate your enthusiasm, but they're fad products. They're heavily promoted, they sell for a few months, then everyone realises they're useless.

MELISSA: The Bun Buster sounds really good to me.

JAKE: Yeah, it would.

MELISSA: What do you mean?

JAKE: Honestly, would you go shopping with elastic bands stretching from your bum to your ankles?

IRENA: Melissa. No woman would be seen dead in them.

MELISSA: They're selling. These things are selling.

JAKE: Melissa, Alex's right. They're junk items.

MELISSA: [*upset*] They're selling!

JAKE: Yeah, for five minutes. And the marketing behind them is putrid. They prey on a population so terrified that they're not physically perfect that they'd buy anything. It's sick.

MELISSA: Shoot me for opening my mouth.

ALEX: No one's going to shoot you around here for coming up with new ideas, Melissa, and I do think there's one area where we should be thinking about expanding our product range. We need a good compact home gym on our product list.

JAKE: That's weird. It's exactly what *I* was going to suggest.

Some time the next day. JAKE *approaches* ALEX.

Sorry, I might've been a bit too tough on Melissa yesterday.

ALEX: If you hadn't said it, I would've had to.

JAKE: [*laughing*] I couldn't believe I was hearing it. What was it, a thigh and buns buster?

ALEX: It's her first job.

JAKE: Yeah.

ALEX: On the home gym, Geoff in production is really good on ergonomics, and Branco's very good on production costs.

JAKE: I've spoken to Geoff already.

ALEX: Good. I should warn you. It's going to have to be a top design to get this past our CEO Douglas. He works on the principle that if he didn't think of it, it's not worth doing.

JAKE: [*nodding*] One of those. Ah, Irena.

ALEX: Yes.

JAKE: What exactly does she do?

ALEX: Collates the product feedback information and writes reports.

JAKE: That's it?

ALEX: [*nodding*] And it arrives at the speed of a very slow snail.

JAKE: [*pointing to Douglas's office on the next floor*] But nobody can say a word against her.

ALEX: Precisely.

Some time later. MELISSA *approaches* JAKE.

MELISSA: Jake.

JAKE: Yes, Melissa?

MELISSA: I've got a problem that I need you to help me solve.

JAKE: Fire away.

MELISSA: When you put down my ideas yesterday, I was really hurt. We have to work together on this project, so if you feel the need to criticise me in public again, do you think you can find a way to do it without causing me to be so upset?

JAKE: [*grinning*] You've done one of those courses, haven't you? 'Jake, I have a problem. When you called me a slack-arsed, cushion-brained loser, I did feel it was a tiny bit harsh.'

He laughs.

MELISSA: Fine.

She turns and starts to walk off. He goes after her.

JAKE: Melissa, just say, 'What you did yesterday gave me the shits. Don't do it again.'

MELISSA: You were the one who encouraged me to talk about those products then you turn around at the meeting and ridicule me.

JAKE: Sorry. If I think something's a bad idea in terms of product outcome, I speak up.

MELISSA: Even if it makes a fool of me?

JAKE: Don't be so sensitive.

MELISSA: If you're going to publicly humiliate me, how are we ever going to work together?!

JAKE: I'm sure we're going to work together very well.

MELISSA: Why?

JAKE: Because it's in both our interests. If we come up with a new product that makes a motza for the company, we're both going to benefit.

MELISSA: There's three of us.

JAKE: Forget Irena. The only reason she's still got a job is…

He points upwards to Douglas's office.

Some time later. JAKE *approaches* IRENA. *She looks up with a weary grimace.*

IRENA: Thank God you put that idiot in her place yesterday. Who in their right minds would employ her?

JAKE: [*shrugging*] Alex must have wanted her.

IRENA: God knows why.

JAKE: I hear you're a calming influence around here.

IRENA: You don't need to dart around like a fire ant to get things done.

JAKE: Absolutely. This new home gym—

IRENA: I can't help with the design.

JAKE: That's my area. All I need to know is whether you'll be available for research now and then.

IRENA: I've got a pretty heavy workload.

JAKE: The time demands should be pretty minimal.

IRENA: Frankly it's a waste of time. The Yanks have perfected the home gym. You won't be able to come up with one as good as theirs.

JAKE: We can try.

IRENA: Even if you do, my uncle'll knock it stone dead. If he doesn't think of something it doesn't happen.

JAKE: Maybe you can work on him.

IRENA: I'm far too far down the food chain for my opinions to count. Despite what Alex thinks, I don't say ten words to him a year. If you want to get his ear get to his PA Francine. She runs everything.

> JAKE *looks at her and nods.*

> *Some time later.* JAKE *has collared* FRANCINE.

JAKE: Francine?

FRANCINE: Yes.

JAKE: Sorry to grab you when you're obviously on your way somewhere, but I'm Jake Jenino, just been appointed to the product development department. They tell me that you're the person who really runs things here, so I thought I'd better say hello as quickly as possible.

FRANCINE: [*smiling*] That's very kind of you, but you've been misinformed. There's only one person who runs things around here and he's my boss.

Rory Williamson as Jake and Melissa Gray as Irena in the 2005 Ensemble Theatre production. (Photo: Steve Lunam)

JAKE: Ah, but behind every successful CEO there's a brilliant PA.

FRANCINE: Now you're just trying to flatter me.

JAKE: Not at all. Three different people have told me that with your capabilities you could step straight out of a support role into top management.

FRANCINE: [*smiling*] What do you want?

JAKE: Nothing at all. I just wanted to make myself known.

FRANCINE: What's Alex got you doing?

JAKE: I'm heading up the design team for a new product.

FRANCINE: Really. Totally new?

JAKE: Yeah, something I suggested that I think could do very well for us. I had quite a success with a radical new pump design at Blackwells. Anyrate, I can see I'm keeping you.

FRANCINE: Busy, busy. Never stops.

> *She turns to move away.*

JAKE: Oh, look. Just one thing. In the other places I've worked I've found it's not much use trying to do something new unless you can get the CEO totally onside.

FRANCINE: That's pretty universal.

JAKE: [*nodding*] I thought it might be wise if I filled Douglas in on progress.

FRANCINE: I assume Alex will be doing that.

JAKE: Apparently Alex likes to keep things close to his chest. My philosophy is to keep your CEO informed every step of the way.

FRANCINE: How would Alex feel if you bypassed him?

JAKE: Yes, you're right. I guess I was thinking how sad it'd be if a great new product never saw the light of day because lines of communication weren't kept open.

FRANCINE: Bypassing your immediate superior is a tricky area.

JAKE: You're right. It's been a pleasure meeting you and I'm sure the rumours are true.

FRANCINE: What rumours?

JAKE: The joint would grind to a standstill without you.

FRANCINE: Been nice meeting you, Jake.

> *He smiles at her and goes. She watches him.*

A little later. FRANCINE *is talking to* DOUGLAS.

You hired yourself a real go-getter in young Jake Jenino.

DOUGLAS: You think?

FRANCINE: Quite an operator.

DOUGLAS: [*nodding*] He's got energy, drive and commitment. Leapt out at you in the interview. Not that Alex could see it.

FRANCINE: Alex wasn't enthusiastic?

DOUGLAS: Practically had to force him on him.

FRANCINE: I think he's someone to watch.

DOUGLAS: So do I. So do I.

> *Some weeks later.* MELISSA *has a folder and papers spread out that she's explaining to* JAKE.

MELISSA: I've surveyed every model that's presently being manufactured worldwide and what we're going to—

JAKE: [*looking at the mass of data*] You *have* been working.

MELISSA: I'm actually getting pretty enthused. I think we can do something world-class. I really do. We're going to have to have at least 200 kilos resistance, a built-in adjustable pulley system, a lat tower, a—

JAKE: A what?

MELISSA: The tall tower at the back here to allow users to their build back and shoulder muscles, and we'll need a leg-curl station which will need to be gym quality—

JAKE: Leg curl?

MELISSA: [*illustrating with leg movements*] Develop your legs. And we'll need a squat station.

JAKE: Sounds obscene.

MELISSA: [*illustrating the squat movement*] Builds your glutes, hamstrings and quads.

JAKE: [*looking at one of the illustrations*] Who designed this gear? The Marquis de Sade?

MELISSA: Ours'll look better. I've actually done some preliminary sketches.

JAKE: [*looking*] Looks good. Sorry I haven't been doing much. I've had a heap of other stuff.

MELISSA: You've done more than Irena.

JAKE: What about that woman. She does bloody nothing.

MELISSA: I've been worried about it. I was thinking that if we sort of included her more, asked her opinion, she might come on board.

JAKE: You're welcome to try.

> MELISSA *looks thoughtful and gathers up her material. She approaches* IRENA.

MELISSA: Hi.

IRENA: Hello.

MELISSA: Thought I'd let you see where we're up to to see if you had any thoughts or inputs.

IRENA: On the home gym?

MELISSA: Here's a few preliminary sketches. I've got to run them past ergonomics and production so they're just rough stabs.

IRENA: [*looking*] Migod. Sorry, but there's no way you'd get me putting one of those in my living room.

MELISSA: Most new homes are being designed with an exercise room.

IRENA: Well, pity about me and my one-bedroom flat.

MELISSA: People like us can join a gym.

IRENA: What's the use of designing something whose only market is the super rich?

MELISSA: It won't be. In the last twenty years the floor space of the average home has almost doubled. Even in blue-collar suburbs.

IRENA: If I had space to burn there's still no way I'd have one of those in my house. [*Sighing*] But then again, who cares about aesthetics these days.

MELISSA: [*trying to contain her anger*] Irena, I've got a problem that you might be able to help me with.

IRENA: Actually I'm pretty flat-chat at the moment.

MELISSA: Irena, you're supposed to be part of a team. When you don't turn up at meetings, or even enquire about our project, I feel you have a negative attitude towards it, or possibly me. Is there something we can do to address this problem?

IRENA: You've done one of those courses.

MELISSA: Irena, we have a problem here. I'm trying to address it.

IRENA: Melissa, frankly I *have* got a negative attitude to the project. I can't think of anything worse than icky, sweaty bodies pounding away on hideous metal frames, stinking up houses. I'm sorry, but I have other priorities in life.

MELISSA: Irena, when you talk like that I feel puzzled. You're—

IRENA: Melissa, would you stop doing that 'When you... I feel' stuff.

MELISSA: [*suddenly losing her cool*] You're paid by a firm that makes exercise machines. If they make you puke, wouldn't it be better to work elsewhere?

IRENA: One does not have to love income tax to work for the Department of Taxation. I do my job, I get paid, and use the money to enjoy the rest of my life. It's what the vast majority of the population is forced to do. If you get off on tubular steel monstrosities then good luck to you.

MELISSA: I enjoy solving difficult problems creatively.

IRENA: Then maybe you should get on with it and leave me to do my work.

MELISSA: [*shocked, almost in tears*] I came here to try and—

IRENA: I'm not a team-type person, Melissa. I have two cats and spend four weeks in Europe a year, which I love. That's my life.

> *She turns back to her work.* MELISSA *stares at her for a second or two, then leaves.* JAKE *comes in, looking at the departing* MELISSA.

Little Miss Gush.

JAKE: [*laughing*] She's certainly full-on.

IRENA: She's orgasmic about this bloody machine she's designing.

JAKE: She's said *she's* designing it?

IRENA: Yes.

JAKE: All she does is run around and make noises. All the thinking is coming from here.

> *He taps his head.*

IRENA: Don't waste your time. It's never going to happen.

JAKE: Don't be too sure. When we do the presentation to Alex, do you want to be part of it?

IRENA: I won't know a thing.

JAKE: Okay. I'll just say that you helped with our research.

IRENA: Melissa will say I didn't.

JAKE: I'll say you did special work for me.

IRENA: Doing me favours won't help you with Douglas. I told you, we never talk.

JAKE: Irena, I was trying to help you cover your arse.

IRENA: Why?

JAKE: I prefer to work in a place where people help each other, not whack knives in each other's backs.

IRENA: [*looking at him*] Okay. Thanks.

> JAKE *looks at her and nods. He leaves.*
>
> *Some time later.* JAKE *and* MELISSA *are facing* ALEX.

ALEX: Is Irena coming?

JAKE: No.

ALEX: Why not?

MELISSA: Because she's contributed nothing.

ALEX: Is that true?

JAKE: [*shrugging*] I hate to point the finger, but yes.

ALEX: God, that makes me angry.

JAKE: I guess we all have to live with it.

ALEX: So. Who's going to tell me precisely where we're at.

MELISSA: Well, I've done quite a—

JAKE: As team leader, maybe… Melissa? I could just start the ball rolling and turn it over to you?

MELISSA: [*hesitating*] Yeah. Sure.

JAKE: We've surveyed every model that's presently being manufactured worldwide and my feeling is that we can do something world-class. We're just at the feasibility stage right now and precise costings have yet to be done but, by God, it's lookin' good. Our basic design is going to have to have at least 200 kilos resistance, a built-in adjustable pulley system, a lat tower, a—

ALEX: A what?

MELISSA: That's the tall tower—

JAKE: [*showing* ALEX] —here at the back to allow the building of back and shoulder muscles, and we'll need a gym-quality leg-curl station which will need to be gym quality—

ALEX: That's for—

MELISSA: [*illustrating with leg movements*] Develops your legs.

JAKE: Develops your legs and of course we'll need a squat station.

ALEX: Which—

MELISSA: Builds your—

JAKE: —glutes, hamstrings and quads.

MELISSA: I've been checking with patents and—

JAKE: There are so many design variants that patent copyright doesn't seem as if it's going to be a problem.

MELISSA: I'm working closely with Geoff on the ergonomic design—

JAKE: He's a real gem.

ALEX: Good, isn't he?

JAKE: Really knows his stuff. We're getting near the point where Branco will be able to draw up some preliminary design alternatives.

ALEX: I'm impressed. It feels like you guys are going for this full throttle.

MELISSA: I know this country still has a mindset—

JAKE: Full throttle and then some.

MELISSA: —a mindset that we can't compete with the US, but I feel just so confident that we can do this.

JAKE: Of course we can compete. In fact there's an advantage in sitting back here, looking at what they can do, and knocking them off. We can import this right into their home market.

ALEX: Hang on there. Look, I really love hearing this enthusiasm, but the crucial factor is costing. Until we've got really accurate dollar estimates we can't even begin to think about production.

JAKE: Absolutely. As soon as we get some designs off the drawing board we'll get a thorough costing underway.

ALEX: I'm impressed. Both of you. Well done.

> JAKE *smiles broadly.* MELISSA*'s smile is a lot more muted.*

> *Later.* MELISSA *confronts* JAKE.

MELISSA: Jake, I'm starting to have a real problem with our working relationship. When you charged in and took credit for all the work I'd done I felt really, really angry.

JAKE: [*his charm disappearing*] Melissa, I'm getting really fed up with this. Did you read the names on the cover of the report? You and I. Both of us. I'm the more senior employee so I presented our joint work. Joint work.

MELISSA: The truth is it wasn't joint work, Jake. I did eighty percent of it. Maybe more.

JAKE: Melissa, you did the research. I pulled it together.

MELISSA: Jake, you rewrote what I'd already written. And yeah, the style was a bit punchier than mine but it was just a rewrite.

JAKE: Just? Have you any understanding of just how important presentation and persuasion are in an exercise like this? Hugely important. Nobody reads detail, they just go with the impact, and that's what I'm good at. Impact.

MELISSA: Well, you know something? I've done all I'm going to do. You handle the costing. I've got plenty of other work.

JAKE: You're withdrawing from the project? Is that what you're saying?

MELISSA: No. I've just done my share. And I'll expect full credit when you submit it.

JAKE: What a cop-out.

MELISSA: I'm not working myself ragged for your glory, Jake. Do some work yourself for a change.

JAKE *catches up to* FRANCINE.

JAKE: Francine.

FRANCINE: Jake. How are you?

JAKE: Doing fine, doing fine.

FRANCINE: So I hear. Good reports.

JAKE: Yeah?

FRANCINE: Rising star, so they say.

JAKE: I was reading an interview with Douglas some time back when he said he very much believed in mentorship.

FRANCINE: It's one of the things CEOs say when they're interviewed.

JAKE: He doesn't believe in it?

FRANCINE: He may very well. He just hasn't *done* it.

JAKE: It's something that was drilled into us in our MBA. Quality mentorship is vital for executive development.

FRANCINE: This isn't a way to get to Douglas about the new project?

JAKE: Not at all. I really took your point on that one. The essence of mentorship is that you don't discuss day-to-day ongoing concerns. Only general business principles.

FRANCINE: Right.

JAKE: I'd just love to hear him talk about how he made his way up the ladder. What to do and what not to do. Purely a mentor situation.

FRANCINE: I could ask him.

JAKE: I'd be really, really grateful. He's got a lifetime of wisdom that I'd like to tap into.

> *Some time later in a city bar.* DOUGLAS *takes a sip of his whisky and looks at* JAKE.

DOUGLAS: I'm all in favour of mentorship, but it's got to be someone with real potential. From what I'm hearing around the place you're making your mark.

JAKE: Thank you, sir.

DOUGLAS: Douglas. But I don't like, repeat, don't like Doug. Douglas.

> JAKE *nods.*

It's a tricky business, mentoring. Really worth doing, but tricky. It has to be a private thing. If section heads get to hear I'm talking with their staff, they get nervous.

JAKE: I understand.

DOUGLAS: And if staff on the same level get to hear, they get envious and insecure.

JAKE: I understand.

DOUGLAS: So there have to be ground rules. No discussion of company matters or personalities. I talk about what I've learned over the years so a bright, young up-and-comer like you doesn't have to re-invent the wheel. That's the essence of mentoring. The tribal elders passing on their wisdom.

JAKE: I really appreciate it.

DOUGLAS: There have been staff I've mentored who thought it was a good opportunity to white-ant others in the organisation, but that kind of behaviour doesn't impress me.

JAKE: Or me.

DOUGLAS: Good. Any questions? Feel free to fire them at me.

JAKE: Great managers. Are they made or born?

DOUGLAS: That's a big one. Both. Some people are born wusses. Scared of their shadows. Some people are born totally insensitive. Can't pick up the vibe. Some people are born without presence or charisma. A major defect like that and all the experience in the world isn't going to help you.

JAKE: Right.

DOUGLAS: The qualities you do need to learn? In a word? Vision. You have to be able to look ahead. See the shape of things to come. If you can't see beyond tomorrow in this cut-throat world, you're dead.

JAKE: Vision.

DOUGLAS: The single most important quality. The next most important, and this is going to surprise you, is humility. Ego's fine. We've all got to have some, but if you've got a monster ego, all you want to hear is praise and you don't listen for the warning signs. A successful manager *must* listen.

JAKE: I can really— [relate to that, in fact…]

DOUGLAS: Must listen. So vital. The manager who surrounds himself with ingratiators and arselickers is doomed.

JAKE: That makes— [so much sense, in fact…]

DOUGLAS: Listen, and listen hard and you're halfway to becoming a good manager. Third most important quality. Resilience. There are times when you've got to be like tempered steel, because disasters are part of the landscape. But you don't view them as disasters. Because they're not. They're opportunities.

JAKE: [*nodding*] Opportunities.

DOUGLAS: Opportunities. And the final thing is building great teams. A great manager is one who doesn't micromanage. Doesn't look over the shoulder of his lieutenants when they're working. Lets them breathe, lets them grow in stature, lets them learn. The great Chinese

General Lao-Tzu worked it all out thousands of years ago. He said that the worst type of leader is one people fear. The next worse is the one people love. The best kind of leader is the one whose subordinates say when the job is done, 'We did it ourselves'. That's the kind of leader I like to think I am.

JAKE: That's the kind of leader— [everyone says you are...]

DOUGLAS: At least that's the ideal. However, when you've got subordinates who start wheeling off in crazy directions, you've got to step in. Can't let a good firm hit the rocks, simply because the CEO's too gutless to intervene when all his warning bells are ringing. I believe you're working on a totally new product line.

JAKE: Yeah. In fact it's looking very exciting at this stage. I think we can probably come up with a world-standard piece of apparatus.

DOUGLAS: I don't want to talk shop, but I have to say I'm surprised no one's keeping me up to speed on this development.

JAKE: Alex apparently believes in having everything in place before coming to you. Personally— No sorry. This isn't the place to discuss it.

DOUGLAS: Personally what?

JAKE: Personally I believe it's better to have your thoughts and wisdom at an early stage, but that's just my take.

DOUGLAS: A pretty sensible take, I have to say. Alex has this odd mentality. It's his little baby and he doesn't want to share it. If he'd have the sense to come to me I could save everyone a lot of time and effort because it's never going to fly.

JAKE: Really? I really feel we're getting somewhere.

DOUGLAS: The project's a waste of time. An utter and total waste of time.

JAKE: Really?

DOUGLAS: You want to know why? This is why. Weight training is dead. No one but mentally retarded beefcake guys want to bulk up any more. And most of those go to gyms where they can watch each other grunt. Everyone knows that for long life and fitness aerobic workouts are what you need. Lung and heart fitness, not biceps. Isn't anyone in your department doing any research?

JAKE: Funnily enough I seem to be the only one who's picked up on the decline in weight training, but Alex is really keen so I felt I better keep my mouth shut.

DOUGLAS: Don't *ever* keep your mouth shut if you're on a boat ride to nowhere. Because my opinion was never asked for I'm left with two options. I let Alex hang himself or I intervene. Now I'm breaking the rules discussing this, but frankly, I don't want a smart young lad like you to go down with the sinking ship.

JAKE: Alex has been so adamant that this is *the* thing. I guess I've felt I had no choice.

DOUGLAS: You've always got choices, Jake. Sometimes they're tough ones, but you've always got 'em. Look, Alex is very smart and very knowledgeable and very hard working, but frankly he lacks the vision thing.

JAKE: It has been worrying me... Douglas.

DOUGLAS: Dogged, tenacious, but he can't inspire.

JAKE: All the effort in the world is wasted if you can't see the wood for the trees.

DOUGLAS: Alex can't see the wood *or* the trees. If he had vision he'd know aerobic devices are the future. There's some wonderful new stuff coming out of the States. My son's a top-ranking triathlon athlete

Henri Szeps (left) as Douglas and Rory Williamson as Jake in the 2005 Ensemble Theatre production. (Photo: Steve Lunam)

and he wears Power Chutes and swears by them. They're a small parachute—

JAKE: I know all about them. I raised them and a lot of other innovative products but got nowhere.

DOUGLAS: Alex. See, no vision. They've increased Grant's endurance enormously he said. You told Alex about them?

JAKE: Yeah. But I have to say I didn't get much of a hearing.

DOUGLAS: Jake, if you've got a boss who's deficient in the vision thing, you have to persist. There's this brilliant machine they've just brought out that enables you to simulate swimming without even getting wet.

JAKE: The Vassar Trainer.

DOUGLAS: The Vassar Trainer.

JAKE: I brought that up too. He said if people want to swim they can go to a pool.

DOUGLAS: Who the hell's got time to traipse off to the local pool?

JAKE: And if you do it's jam-packed with lap nazis.

DOUGLAS: By and large people *hate* swimming, but they know how good it is for you. That's the brilliance of a machine like that. And my wife Sue Ellen just got back from the States with this device like two giant rubber bands that stretch from your waist to your ankles and give your bum and legs a workout.

JAKE: The Thigh 'N' Buns Buster. Brought that up too.

DOUGLAS: And he wouldn't listen?

JAKE: He feels the home gym is the way to go.

DOUGLAS: He couldn't be more wrong. If we need anything, we need a modern light-framed aerobic machine.

JAKE: I'm so glad to hear you say it.

DOUGLAS: Speak up. Don't waste your career labouring on a dinosaur that's never going to go into production. That other girl who's working with you. Will she back you?

JAKE: Look, she's got her good points, but—no, I don't think so. When Alex says jump, she jumps.

DOUGLAS: My niece Irena?

JAKE: She's with me.

DOUGLAS: Odd girl. Had some qualms about appointing her, I have to say. Her mother's inclined to think she's lesbian.

JAKE: She's calm, steady, but not big on the vision thing.

DOUGLAS: The lesbian thing's not a problem, as long as she fits in.

JAKE: She's fine.

DOUGLAS: Glad to hear it. My sister'd give me hell if I had to fire her.

JAKE: She's fine.

DOUGLAS: Yes. It seems your problems lie elsewhere.

> *Some days later.* ALEX *listens with a deep frown as* JAKE *holds forth.* MELISSA *and* IRENA *are also at the meeting.*

JAKE: Alex, it's given me a lot of sleepless nights, I have to say, but I finally felt I have to speak my mind. The home gym is a dinosaur.

ALEX: Dinosaur?

JAKE: I just feel we could be wasting our time.

MELISSA: Home gyms are still selling very well.

ALEX: If they weren't, I wouldn't've had them on the agenda.

JAKE: I just sense that the future is lightweight aerobic machines and other aerobic devices.

ALEX: You sense?

JAKE: Things like the Power Chute, and the Thigh 'N' Bun Buster. I know I expressed some doubts—

MELISSA: Some doubts?

JAKE: I think I was wrong.

ALEX: What is this, Jake? You want me to pull the plug on months of work and career off in the gimmick direction?

JAKE: I just think lightweight aerobic devices are the future.

ALEX: Where did this sudden new conviction come from?

MELISSA: Jake, you ridiculed me for even suggesting those sort of things.

JAKE: I was wrong and I'm big enough to admit it.

ALEX: Sorry, Jake, but you'll stick to the home gym concept. I want it pushed through to full presentation stage. Costings, market research, production runs, the lot.

JAKE: [*shrugging*] You're the boss.

ALEX: [*tensely*] I'm finding it hard to come to terms with the fact that you're ready to totally junk something you seemed wildly excited about just three days ago.

JAKE: And I was, but I woke up that very night and found myself asking what is it that people really want. And it's simple. They want to live longer and they want to live better. They want healthy hearts and healthy lungs and we should be building machines that help them achieve that. If you choose not to listen then be it on your own head.

>ALEX *stares at him.*

ALEX: [*exploding*] Who the hell do you think you are?!

JAKE: If I'm not allowed to say what I truly believe then I'm happy to resign.

ALEX: Jake, you'll press on with the home gym to full presentation stage, and you'll help him, Melissa. And, Irena, you're supposed to be part of this team, so I'd like to see you genuinely involved.

>*He picks up his papers and storms out of the conference room. The other three look at each other.* MELISSA, *in particular, stares at* JAKE.

JAKE: Okay. You were right. I've changed my mind.

MELISSA: Too late now. It's home gym right through to the end.

>*She gets up and leaves.* IRENA *turns to* JAKE.

IRENA: Good for you. I've always thought it was a dog.

JAKE: Could you do me a big favour?

IRENA: What?

JAKE: Write me an email that sort of recounts what happened here today.

IRENA: What, everything?

JAKE: No no. Just that you thought I was right to say what I did and you were disappointed when Alex blew up at me. I'd like something on record so that when the project goes down the tubes I won't go down with it.

IRENA: I don't like putting things in writing.

JAKE: I won't ever show it to anyone without getting your permission first.

>*Some evenings later.* DOUGLAS *is reading a printout of an email that* JAKE *has given him as he sips his scotch in the city bar he frequents. He looks at* JAKE.

I know I shouldn't be showing it to you.

DOUGLAS: He blew up?

JAKE: Screamed at me in front of everyone.

DOUGLAS: Very poor management technique.

JAKE: I was totally humiliated.

DOUGLAS: Rule number one. Reprimands must be delivered in private.

JAKE: Screamed at me. I thought he was going off his brain.

DOUGLAS: That's worrying. Even more worrying is that he's charging on with this pet white elephant.

JAKE: When you said that one has to speak out, I guess I took it a little too literally.

DOUGLAS: You said what you believed and he should've listened. Every good manager listens. They don't lose it and scream. And then tell you to keep working on a dud project.

JAKE: It just seems such a waste of time and effort.

DOUGLAS: It is.

> *Some time later.* ALEX *faces* DOUGLAS *in Douglas's office.*

ALEX: What's the message I'm being given here? You want me to pull the plug?

DOUGLAS: Yeah, that's essentially what I am saying, Alex.

ALEX: For God's sake, Douglas, it's only a proposal. You were always going to have the final say.

DOUGLAS: Alex, when I saw the words 'Home Gym' on your quarterly report I had a sinking feeling. It's not the way of the future.

ALEX: Have you seen the sales figures?

DOUGLAS: I've seen last year's sales figures. What I've got to do is see the sales figures five years, ten years ahead. And I don't like what I see.

ALEX: No one can forecast—

DOUGLAS: The reason I'm sitting in this chair, Alex, is to make that sort of call. And as long as I'm sitting in this chair, my call is your reality. Lightweight aerobic equipment is the future.

ALEX: Let's get this straight. You're telling me I have to pull the plug before you even see the proposal?

DOUGLAS: Yes. It's a waste of time and effort.

> ALEX *turns furiously and goes.*

A short time later. ALEX *storms into Irena's workspace.*

ALEX: I hope you're glad.

IRENA: About what?

ALEX: Your uncle's stopped our home gym project.

IRENA: It's nothing to do with me.

ALEX: Isn't it?

IRENA: Alex, I haven't spoken one word to my uncle in the last year.

ALEX: How come he was quoting exactly the same language as Jake was in our supposedly confidential meeting? Lightweight aerobic machines are the future, word for word.

IRENA: [*upset*] Credit me with some integrity. It's bad enough that everyone knows he got me my job here, without them thinking I'm his in-house spy.

ALEX: [*to the audience*] I should've been smart enough to work out that the circuit to Douglas was Jake. I guess there was some part of me that wanted to believe that bastardry of that type didn't happen. More fool me. [*To* IRENA, *angrily*] Well, it's bloody suspicious. That's all I can say. Bloody suspicious!

ALEX *leaves.*

Some time later. JAKE *approaches* IRENA *who has been crying.*

JAKE: What's up?

IRENA: Alex just accused me of going behind his back to my uncle on this home gym thing.

JAKE: You're kidding.

IRENA: Douglas apparently hates the idea as much as we do, which means, according to Alex, that I must have primed him.

JAKE: The idea's a dog and Douglas is paid to spot dogs.

IRENA: Precisely! [*She starts to cry.*] Sorry, I hate doing this.

JAKE: Did he lose it again?

IRENA: Shouted at me.

JAKE *shakes his head and lets out a sigh.*

As if my life isn't a big enough mess as it is.

JAKE: What's wrong?

IRENA: Don't ask.

JAKE: No, tell me.

IRENA: The ludicrous thing is that Uncle Douglas and I have probably exchanged less than a hundred words in my life. He thinks I'm weird.

JAKE: Why?

IRENA: He thinks I'm a lesbian.

JAKE: Why would he think that?

IRENA: Probably because I am. You didn't guess?

JAKE: No.

IRENA: The one great thing in my life was my partner Pamela. All year we planned our trip to Italy, practised our Italian and then for four glorious weeks we escaped from... you don't want to hear this.

JAKE: Go on.

IRENA: We escaped from the *horror* of this kind of life and went to Firenze, Sienna, Umbria and felt we were in heaven. And then, stupid me, I introduce her to Denise, an old friend, then suddenly Pamela's being spiteful, sarcastic, vindictive, and I'm getting more and more bewildered and hurt, until I finally work it out.

JAKE: She's giving you the brush-off so she can move in with Denise?

IRENA: [*nodding*] It was just so cruel and hurtful. The way she did it.

JAKE: That's awful. And now this.

IRENA: [*indicating the departed* ALEX] I tried not to cry in front of him, I didn't want to give him the satisfaction, but I'm very vulnerable right now.

JAKE: Don't let him get away with it.

IRENA: [*shaking her head*] I'm not going to go running to Douglas.

JAKE: In general I'd agree, but this time I think you should.

IRENA: No.

JAKE: He's lost it with me. He's lost it with you. It just shouldn't be happening. No department should run on intimidation and fear.

    IRENA *looks at him.*

    Some time later. ALEX *faces* DOUGLAS.

ALEX: [*incredulous*] Anger management?

DOUGLAS: I think you should have some type of counselling, Alex.

ALEX: I lost it briefly.

DOUGLAS: Twice. Look. This company needs you. Couldn't do without you. You're a walking encyclopaedia on the evolution of everything we make. You're totally irreplaceable.

ALEX: For God's sake, I don't need therapy.

DOUGLAS: Not therapy. Just counselling.

ALEX: No.

DOUGLAS: You're not a little concerned?

ALEX: No.

DOUGLAS: Alex. One of your projects is criticised, in my opinion for very good reasons, and you try and stifle dissent by brute force. It's not the way things should happen in a healthy department.

ALEX: I believed in that project. I still do.

DOUGLAS: Well, I think you were wrong, and you should be able to accept that without degenerating into a lather of aggression.

ALEX: A lather of aggression?

DOUGLAS: My niece has never once come to me except this time. That's why I know it's serious.

ALEX: If you tell me the home gym is a dead item, then okay, I move on. But I don't want or need counselling.

DOUGLAS: [*with a sigh*] It was a suggestion, Alex. A well-meaning suggestion out of concern for you, and the company. I'm not about to force you.

ALEX: Everyone loses their temper occasionally.

DOUGLAS: Alex, let's try and be positive. Let's try and put this unfortunate business behind us, and make every post a winner from here on in. Okay?

> ALEX *looks at him, still simmering with resentment, but eventually nods and goes.*

> *Some time later.* ALEX *is addressing* JAKE, IRENA *and* MELISSA.

ALEX: Well, we're back to square one. I remain far from convinced that our home gym was a bad idea, but our CEO has very firmly decided that it is, so let's move on. Lightweight aerobic machines. This is a difficult field because all the successful machines are highly protected by patent, so it appears we have to start totally from scratch. This is the direction you wanted to go in, Jake, so I presume you're brimming with ideas.

JAKE: I've been talking to Geoff and Branco and, yes, this is going to be a difficult one. We do have to start from scratch.

MELISSA: Maybe not. I've been researching and there's a South Australian company that went broke before they could get their design onto the market, but the design itself appears to be a good one. It's a variant on the cross-country skiing machine, but it's original enough to get past our patent worries, and I think they'll sell it to us for a very reasonable royalty cut.

She hands around some sketches. They look at them, nodding with interest.

ALEX: Okay. I've got a heap of other things to do. So I'm going to leave this entirely to you guys. Buy the patent, develop and cost it, and find prospective buyers before you get back to me. This is your baby.

He looks at them. Then leaves.

IRENA: I know my limitations and you know them too, but honestly, if there is anything I can do to help, I'll do it. At the very least I'll document and file all the paperwork.

MELISSA: Thanks, Irena.

JAKE: That'll be a great help.

She leaves. MELISSA *looks at* JAKE.

MELISSA: She's looking like someone who desperately needs Prozac.

JAKE: Leso love angst.

MELISSA: Yeah?

JAKE: Her girlfriend left her for someone slightly less ugly than she is.

MELISSA: Jake, don't be a bastard all the time.

JAKE: Sorry.

MELISSA: When you're in love and the other person couldn't care less, it hurts.

JAKE: Colin Firth's not responding to your vibe?

MELISSA: I wasn't referring to anything here.

JAKE: You're lying. I hear those lovelorn sighs.

MELISSA: I was referring to something that happened to me years ago.

JAKE: Tell me.

MELISSA: I'd feel safer confiding in a public broadcast system.

JAKE: Charming.

MELISSA: You're the one who's charming. When it suits you.

JAKE: If you've finished with the character assassination, can we do something productive?

MELISSA: Okay, I want one thing absolutely straight, right from the start.

JAKE: I know what you're going to say.

MELISSA: Yeah, well you should know what I'm going to say. I'm not going to work my guts out and have you stealing the credit like you tried to do last time.

JAKE: It won't be a problem.

MELISSA: It better not be a problem, because I won't keep quiet about it this time.

JAKE: Okay, I'm sorry. I let my enthusiasm get the better of me, but I understand how you feel and I'll be a lot more careful. And, hey, that was great work to locate this baby. It looks really, really promising.

MELISSA: Let's not get carried away until ergonomics and production have put it through the wringer.

JAKE: Think positive. The good thing about Alex cutting us adrift is that if we pull this off the glory's all ours.

MELISSA: The flip side is that if it's a turkey we go nowhere, so just calm down. Let's just see what Geoff and Branco think.

JAKE: I've got a feeling about this. It's a winner.

MELISSA: Calm down.

JAKE: Melissa, we could be about to become the golden kids.

MELISSA: Just make sure that if we do pull it off you don't try and grab all the credit.

JAKE: Melissa honey. This is *our* project. Watch my lips. Ours.

MELISSA *looks at him, still not totally convinced.*

END OF ACT ONE

ACT TWO

MELISSA *and* JAKE *watch as* ALEX *looks at the paperwork they have put in front of him.*

JAKE: It all stacks up. Everything stacks up. The machine does deliver the fitness benefits, and production can handle it without any major dramas. Consumer research gives it top marks. And the costing is a little higher than we'd hoped but still competitive.

ALEX: It's not a cheap machine.

MELISSA: Check the comparison chart on page seventeen.

> ALEX *turns to the page.* MELISSA *points.*

It's still competitive with any comparable machine.

ALEX: It's not cheap.

JAKE: It's competitive.

ALEX: Let's put that to the test then. If you can get over three thousand orders out there I'll ask Douglas to okay it for production.

MELISSA: Three thousand?

ALEX: We need a run that size to amortise tooling costs.

MELISSA: It's a big ask.

ALEX: If it's as good as you think it is, you should be fine.

JAKE: We'll get them.

ALEX: If you do, then well done. I know I've had reservations about this, but this is good work, and if you can make this happen I'll be giving you all due credit.

JAKE: Appreciate that, Alex. Some bosses would do anything in their power to grab the glory themselves.

ALEX: I don't operate like that, Jake.

MELISSA: Can I get this straight? We have the authority to sign deals with the retailers?

ALEX: As long as production have signed off on cost. It's your baby from go to whoa.

> JAKE *nods.*

Some time later. JAKE *is confronting* MELISSA.

JAKE: Production won't budge?

MELISSA: Geoff says he'll come down one dollar twenty-five per unit and that's absolutely it.

JAKE: That's not going to help.

MELISSA: I know.

JAKE: The truth is, if push came to shove, they could shave fifty dollars off the delivery price. I've done the costing myself. They're just protecting their arse.

MELISSA: Probably right.

JAKE: They're killing us on cost out there. What is it? We've got—

MELISSA: A hundred and thirty-five orders.

JAKE: And we need three bloody thousand. We're never going to get them at that price. Go back to production again.

MELISSA: Geoff is adamant. That's as far as he'll go.

JAKE: [*angry*] Bastard. This whole firm is infected with negativity.

MELISSA: Don't let it get to us. Let's keep hammering at the big outlets until they refuse to let us in the door.

JAKE: Production has sabotaged us and Alex's probably urging them on.

MELISSA: Just keep working at it. [*She looks at her watch.*] Damn. I've got another appointment with the Gymworld guy. Will you phone his secretary and say I'm on the way?

JAKE: How are we doing with K-Mart?

MELISSA: Forget it. Unless we can shave forty dollars off the unit price they told me not to come back.

JAKE: If we could?

MELISSA: Ten thousand units.

JAKE: Ten thousand? For a run that size surely production could get costs down.

MELISSA: Yeah, maybe another two dollars says Geoff.

JAKE: That *has* to be bullshit!

MELISSA: I spent an hour pleading. You try.

JAKE: Ten *thousand*. We'd be flying.

MELISSA: You argue with Geoff. Will you ring that Bunnings guy?

JAKE: What's his number?

MELISSA: Owen Griffith. His details are in my computer. Ring his mobile. Tell him I'm on my way with a new offer, but don't tell him how pathetic it is.

JAKE: K-Mart would take ten thousand?

He thumps the table and swears, enraged. MELISSA *nods in sympathy and leaves.*

Shortly after. JAKE *is sitting at Melissa's workstation with the phone in his hand.*

[*Into the phone*] She's on her way now. She's, er, bringing a new offer. Not huge, but I think you're going to be pleasantly surprised. Yeah. Same to you, buddy.

He hangs up. He's about to get up from the workstation when he looks back at the screen. Something catches his eye. He goes into the email address book and searches. He finds what he's looking for. He hesitates, then impulsively and with energy starts tapping out a message. IRENA *appears.*

IRENA: Why are you working on Melissa's computer?

JAKE: [*startled*] God. You gave me a fright. [*Indicating the machine*] She's racing to make appointments. She wanted me to email some details to buyers I haven't got in my address book.

He positions himself so IRENA *can't see the screen and seems a little tense during the next exchange.*

IRENA: Going to get married, I see?

JAKE: Engaged. How did you know?

IRENA: Your photo was in the social column.

JAKE: I missed that. Where?

IRENA: *Sunday Telegraph*. I cut it out. Do you want it?

JAKE: Thanks. Did I look good?

IRENA: No. She's nice though.

JAKE: Pretty good, eh?

IRENA: And wealthy, wow.

JAKE: Not bad for a boy from Maroubra, eh?

IRENA: You must have something.

JAKE: I'll show you if you like.

IRENA: Wrong girl for that.

JAKE: How's, er, things?

IRENA: Just when you think nothing can get worse, my cat gets cancer. I can pay over a thousand for an op or have her put down. Great choice.

JAKE: Save the cat and get that two-timing girlfriend put down.

IRENA: I finally sent her one hell of a letter.

JAKE: Good for you.

IRENA: Made me feel a little better. I'll let you get on with it.

JAKE: Sorry about your cat.

> IRENA *nods and goes.* JAKE *looks around and resumes frantically tapping out the email.*
>
> *A few days later.* JAKE *appears in front of* MELISSA *in a state of some excitement.*

Sign this.

MELISSA: What is it?

JAKE: We're home, baby. We're home.

MELISSA: What is it?

JAKE: An order from K-Mart. Ten thousand units. For starters. If they move them, the sky's the limit.

MELISSA: My God, how did you do it? I went back to their guy three times and he wouldn't budge.

JAKE: Phoned the guy cold. Thought I'd give it one more try. Hope you don't mind.

MELISSA: Mind? That's brilliant.

JAKE: Just sign the agreement here. Then I will. [*He flips over the page and indicates where.*] It's got to have both of our monikers on it to make it official.

MELISSA: Let me read it first.

JAKE: Sweetness, I have to have this couriered back half an hour ago. All you need to read is the figure 'ten thousand'. And sign.

> *She does so.*

Now me.

> *He scribbles his signature.*

MELISSA: Leave me with a copy.

JAKE: No time now. We'll do that later.

He puts the document in an envelope and seals it.

Oh, look. I've got to get to a production meeting urgently. Can you call the courier and sign for this?

He hands MELISSA *the sealed envelope and goes.*

Some days later. MELISSA *approaches* ALEX. *He looks up.*

ALEX: Thanks for coming.

He hands her a document.

You signed and couriered this delivery contract?

MELISSA: For ten thousand SkiFits? I thought you'd be pretty pleased.

ALEX: Well I was, until I read the delivery price.

MELISSA: It'll be whatever production signed off on.

ALEX: [*shaking his head*] I checked with production. You've committed us to forty dollars less per unit than you were authorised to.

MELISSA: No, that couldn't be right.

ALEX: Melissa, you signed this.

MELISSA: Yes, but I assumed the price was the approved one.

ALEX: You signed this without checking the price?

MELISSA: I assumed Jake had checked the price.

ALEX: It's not Jake's signature that's on here. It's yours.

MELISSA: Jake did the deal.

ALEX: Where's his signature?

MELISSA: He signed it. I saw him. [*She flips the page to check and stares at it.*] I saw him.

ALEX: Melissa, what the hell's going on?

MELISSA: Jake did the deal.

ALEX: And you signed without reading it?

MELISSA: He said it was urgent.

ALEX: And you saw him sign?

MELISSA: Yes.

ALEX: Show me, Melissa. Show me that signature.

MELISSA: It's not there, obviously. This is a photocopy.

ALEX: It's a photocopy of the original.

MELISSA: He signed.

ALEX: Melissa, do you realise the implications? You've just lost the firm four hundred thousand dollars.

MELISSA: Phone the guy at K-Mart. He'll confirm it was Jake made the offer, not me.

ALEX: I have. The offer came via email written on your machine and e-signed by you.

MELISSA: He must've used my machine.

ALEX: Melissa.

MELISSA: Alex, look at me. If I'd made a mistake like this I'd own up. Don't you know me well enough by now? Don't you know him well enough?

ALEX: An email signature, okay, but this is a real signature written by a real pen. And it's your signature on the courier chit.

MELISSA: I signed them both. I'm not denying that. But he signed the contract too.

ALEX: Then where's his signature?

> MELISSA *frowns, then suddenly takes a pen and bends down and dashes off a signature, or appears to. She hands it to* ALEX.

MELISSA: Where's my signature?

ALEX: [*looking at it*] I want to believe you, Melissa. I really do.

MELISSA: He pretended to sign. The pen never touched the paper. Just don't do anything until I get to him. Please. I'll *make* him tell the truth.

ALEX: Straighten this out quickly, Melissa. Dropping four hundred thousand is not going to amuse anyone.

MELISSA: Just let me get to him.

> MELISSA *leaves.*

> *Shortly after.* MELISSA *approaches* JAKE.

JAKE: What's wrong?

MELISSA: What's wrong? You quoted forty dollars under the approved price on my computer and put my signature to it. Then got me to sign the contract and organise the courier.

JAKE: For God's sake, keep it down. Don't panic.

MELISSA: Don't panic? We're looking at a huge loss. No wonder K-Mart signed. We're going to make a four hundred thousand dollar loss.

JAKE: That's bullshit. When production realise that they have to come in at that price they'll do it, believe me. Happens all the time. And even if production makes a slight loss on the first run, we're out there in the market and we can bump up the prices then. Hold your nerve and don't panic. We're still going to be heroes around here.

MELISSA: Alex knows. He checked with production.

JAKE: Shit!

MELISSA: Go to him right now and tell him the truth. This is all your doing.

JAKE: Okay. Okay. I will.

MELISSA: Right now.

JAKE: Okay. Okay. I'll tell him that if he holds his nerve, this thing'll be fine.

> MELISSA *stands there and watches as he goes. As he makes his way there he runs into* FRANCINE.

FRANCINE: Jake, I was looking for you.

JAKE: Francine.

FRANCINE: I've just had Geoff from production in my office breathing fire. He says someone in your department contracted to supply machines at way under cost. He's in with Douglas now. I just thought I better warn you.

JAKE: Oh God. Poor bloody Melissa.

FRANCINE: The girl you work with?

JAKE: The girl I *try* and work with.

FRANCINE: Is she difficult?

JAKE: Try pig-headed, headstrong and irresponsible.

FRANCINE: Seems like she's landed herself in a mess this time.

JAKE: Does things on impulse then tries to lie her way out of it.

FRANCINE: Some people just seem to let their ambition override their common sense.

JAKE: [*nodding*] I actually warned her about this and she still went ahead.

FRANCINE: Judging from how mad Geoff was, I think she's in for a rocky time.

JAKE: Thanks for letting me know, Francine. I really appreciate it.

He nods at her gravely and leaves.

A little while later. JAKE *and* MELISSA *and* ALEX *are being interviewed by* DOUGLAS, *who isn't happy.*

DOUGLAS: Do you know how it feels to me, as CEO, to be legally committed to supply a product at a massive loss? Alex, the buck stops with you. How could you let it happen?

ALEX: I accept responsibility. The only upside is that if SkiFit is a big success we'll make the money back and eventually go into profit.

DOUGLAS: And if it's not a big success?

ALEX: Then we've done a lot of dough.

DOUGLAS: You're in charge of that department. How could you let this happen?

ALEX: I gave the team autonomy. I wanted them to take responsibility. But one of the guiding parameters was that they were not to quote below the figure production set. I assumed they'd respect that instruction.

DOUGLAS: Melissa. Can you explain why you did this?

MELISSA: The whole deal was done by Jake.

DOUGLAS: Then where's his signature?

MELISSA: He pretended to sign but he didn't.

DOUGLAS: And the email from your computer?

MELISSA: He must have used it when I was out.

DOUGLAS: And your signature on the courier chit?

MELISSA: He asked me to send it.

DOUGLAS: So he meticulously planned this all like a thoroughgoing criminal?

MELISSA: Yes.

DOUGLAS: That's pretty far-fetched.

JAKE: It's a total lie.

MELISSA: He knew it was risky, so he wanted someone else to take the blame if it backfired.

JAKE: Melissa, for your own sake tell the truth.

DOUGLAS: What's your version, Jake?

JAKE: She came to me and told me what she'd done and asked me to sign. I was appalled. I told her she was crazy. I told her there was no *way* I would sign. She yelled at me and told me I was a wimp and said she'd sign. I tried to make her see reason. I tried to point out the risks, but she had this idea in her head that production departments inflate costs to protect their own arses—

MELISSA: Jake, *you* said that!

JAKE: And she said that if production knew they had to produce at that cost they'd pretty soon find a way to do it.

MELISSA: *You* said that!

JAKE: I said she might be right, but it was one hell of a risk and there was no way I was going to be part of it.

ALEX: Why didn't you report her behaviour to me?

MELISSA: It's a total lie! He's turned everything around.

ALEX: Jake, why didn't you report her behaviour to me?

From left: Henri Szeps as Douglas, Rory Williamson as Jake, Michael Ross as Alex and Katrina Milosevic as Melissa in the 2005 Ensemble Theatre production. (Photo: Steve Lunam)

JAKE: I guess it's just not in my nature to dob someone in, but you're right, I should've. Although when I saw her sign the contract and race off to courier it, I knew there was nothing I could really do. I guess by that stage I was praying that she was right and that production could do it at that cost.

MELISSA: Everything he said is a lie.

DOUGLAS: Melissa, please. It was your email and your signature is on both documents.

MELISSA: He must have sent the email from my computer when I wasn't there.

DOUGLAS: Alex, my information is that Melissa is known to be headstrong and irresponsible.

ALEX: That's not true. She's energetic, sure and motivated.

DOUGLAS: And does she have a habit of covering up her mistakes?

ALEX: Not to my knowledge.

DOUGLAS: It's certainly something I've heard.

MELISSA: [*distraught*] That's not true.

DOUGLAS: I'd like to have a word to Alex. Could you both go to your workstations. I'll call you if you're needed.

MELISSA: [*very upset*] Alex, you know I'm telling the truth.

DOUGLAS: Melissa, if you could just leave us for a little while. Jake.

> JAKE *nods and leaves.* MELISSA *turns to go, but turns back.*

MELISSA: Alex, you know what he's like.

ALEX: Melissa, I'll talk to you shortly.

> *She goes.* DOUGLAS *turns to* ALEX.

DOUGLAS: So what do you make of this?

ALEX: I tend to believe her.

DOUGLAS: The email? The signature?

ALEX: I know. It's just not like her to lie.

DOUGLAS: You're saying that not only is Jake lying, but he's also extraordinarily devious?

ALEX: It's not like her to lie.

DOUGLAS: It is for Jake?

ALEX: More so.

DOUGLAS: Has he lied to you? Can you give me instances?

ALEX: No. He's clever.

DOUGLAS: I think the girl's been caught out and is terrified. You *buy* that phoney story of him faking the signature?

ALEX: I tend to. Yes.

DOUGLAS: Could it be just possible that you've had it in for Jake ever since I recommended hiring him?

ALEX: He's intelligent, plausible, and all that—

DOUGLAS: You were biased against that lad right from the start and you know why?

ALEX: I'm not biased.

DOUGLAS: Could it be that Jake is a bit too impressive for you to feel comfortable?

ALEX: Are you serious?

DOUGLAS: Young bull, old bull. It happens.

ALEX: Douglas, Jake has got years to go before he knows this business. I'm not the slightest bit threatened.

DOUGLAS: I think he's a fast learner and top-drawer material. And I think that Melissa is patently at fault, and that she's got to be dismissed immediately, and I want you to do it.

ALEX: I think she's telling the truth.

DOUGLAS: Well I don't. And if you don't tell her to clear her desk, then I will.

Down below, JAKE *has taken* IRENA *aside.*

JAKE: Everything's gone to shit, Irena. Melissa quoted forty dollars under what we were authorised to quote and the firm's going to lose four hundred thousand.

IRENA: You're kidding.

JAKE: And she's flailing around trying to blame everyone but herself. Notably me.

IRENA: You didn't know she made the offer?

JAKE: [*shaking his head*] Typical headstrong Melissa.

IRENA: I can't believe she didn't check with you.

JAKE: It's her signature on the bloody deal.

IRENA: Then how can she claim it's you?

JAKE: She's got some cock-and-bull story that I pretended to sign but didn't really. Jesus!

IRENA: Sounds desperate.

JAKE: Totally desperate. She made the offer by email and she's trying to say I sent it from her computer.

IRENA: Her computer?

JAKE: I've never been near her bloody computer.

> IRENA *looks at him. He frowns.*

Oh yes, that one time when we were talking about your cat. All I was doing was sending a message she asked me to send to Gymworld.

IRENA: Then it'll be there on her hard drive.

JAKE: I didn't send it in the end. I thought it'd be quicker to phone.

IRENA: Right.

JAKE: I'd forgotten about that. If Alex asks…

IRENA: I've got to tell the truth.

JAKE: She asked me to send a message. I swear.

IRENA: Jake, I can't lie.

JAKE: Great. Alex will use it to get her off the hook and there'll be a cloud over me when I'm totally innocent.

IRENA: I can't lie.

JAKE: Can I tell you something? I wouldn't normally do this, but you should know the truth about the person whose career you'll be saving.

IRENA: What?

JAKE: Remember when you told me about the rough time you'd been having?

IRENA: Yes.

JAKE: She asked why you were going round looking like someone who desperately needs Prozac. So I told her and she said, 'Oh, for God's sake. All that drama over a little bit of leso love angst.' And I said, 'Come on, when you lose someone you love it really hurts'. And she said, 'Why doesn't she face up to reality? Her girlfriend left her for someone slightly less ugly than she is.' You want to have someone like that around here for the next ten years?

> IRENA *says nothing.*

I didn't send that offer, Irena. I just wouldn't do something like that.

IRENA *says nothing.*

Some time later. IRENA *is approached by* ALEX.

ALEX: Irena, this is all very distressing. I know you had nothing to do with it.

IRENA: No, I didn't.

ALEX: It's a case of two competing stories.

IRENA: Yes.

ALEX: You work opposite Melissa. The email was sent at two forty-three last Tuesday. Melissa thinks she'd left for Gymworld, but can't be sure. Did you see anyone working on Melissa's computer at around that time?

IRENA: [*hesitant*] No.

ALEX: Could it have happened when you were away from your workstation?

IRENA: I'm at my workstation most of the time.

ALEX *nods.*

ALEX: Right.

Some time later. ALEX *sits with* MELISSA *in a café after work.* MELISSA *stares straight ahead. Eyes expressionless.*

Melissa, as a matter of urgency I've lined up an appointment tomorrow with a counsellor who I'm told is very good.

MELISSA: I don't want a counsellor.

ALEX: Melissa, you've been through a major trauma.

MELISSA: Can't anyone else see what slime he is?

ALEX: He's clever. He's covered all of his tracks.

MELISSA: Yes, but can't anyone *see*?

ALEX: We've got no evidence. I'm fairly sure Irena did see him at your computer, but she's never going to admit it.

MELISSA: She loathes me. I have no idea why. I can't recall ever doing anything to deserve it.

ALEX: God knows what goes on in her head.

MELISSA: How can someone lie without one flicker of guilt?

ALEX: They're plenty of them and they tend to do just fine.

MELISSA: What's the use then? If I get another job there'll just be another Jake, then another, then another.

ALEX: Am I like Jake?

MELISSA: No, no. Absolutely not.

ALEX: And neither are you. And there are more of us than there are of Jake. If there weren't, the world would be stuffed.

MELISSA: It is.

ALEX: He'll be found out eventually.

MELISSA: Wishful thinking.

ALEX: He'll be found out.

MELISSA: You know why I wanted this job so badly?

ALEX: Why?

MELISSA: You. In the interview. I just felt you were someone I could... work for.

ALEX: I wanted you to work for me.

> MELISSA *looks at him. Looks away, then looks back.*

MELISSA: Keep me company. Come home with me.

> ALEX *stares at her.*

I'm feeling really low.

ALEX: Melissa, it wouldn't be a good idea.

MELISSA: Why?

ALEX: Your judgement's not going to be great right now. You're very vulnerable.

MELISSA: This is not spur of the moment, Alex. I've felt strongly about you for quite a while. As if you didn't know.

ALEX: Actually, I didn't.

MELISSA: Jake, with his laser-like capacity for spotting human weakness, knew straight away. Which is probably why I made every effort not to show it.

ALEX: Melissa, this is the worst possible time.

MELISSA: Why?

ALEX: You've just been through a major trauma.

MELISSA: I guess the truth is you've never found me the slightest bit attractive in any case.

*Katrina Milosevic as Melissa and Michael Ross as Alex in the 2005
Ensemble Theatre production. (Photo: Steve Lunam)*

ALEX: That's not true.

MELISSA: I'm overweight, I'm daggy—

ALEX: Melissa, you're very, very special. Right from the time you walked into the interview room I've thought that.

MELISSA: Then, for God's sake, come home with me.

ALEX: Melissa, it's not the time. You go to the counsellor first thing in the morning. That's first priority.

MELISSA: It's not mine.

ALEX: Well I'm sorry, it's mine. [*To the audience*] I did everything by the book. What I said was right, principled, ethical, moral and totally, totally stupid. Which is probably why I was so angry the next day.

ALEX *confronts* JAKE *and* DOUGLAS.

DOUGLAS: Just how bad is she?

ALEX: They've pumped her stomach. With luck she'll probably survive.

DOUGLAS: So what are you trying to do? Sheet the blame home on me?

ALEX: She tried to kill herself because she was *totally* innocent.

DOUGLAS: Alex, you're not being rational. The evidence was overwhelming.

ALEX: Jake typed the email on her machine. Jake faked doing his signature. Jake asked her to call the courier.

JAKE: If you weren't so obviously emotional I'd take you to court over those accusations.

ALEX: Yeah, you probably would.

JAKE: Alex, whether you realise it or not, you're emotionally involved with her.

ALEX: Don't try that.

JAKE: She's been besotted with you from the minute she arrived here. And whether you did anything about it or not, I've always had the feeling it's been reciprocated.

DOUGLAS: Is this true?

ALEX: Of course it's not bloody true.

JAKE: I'm not saying you did anything about it. Just that your feelings have distorted your judgement.

ALEX: You're the reason she nearly died last night and you know it.

JAKE: [*angry*] That's absolute bullshit!

DOUGLAS: Cut it out, Alex. That's bloody libellous.

ALEX: He's a cold, manipulative shit. Which you'll find out sooner or later.

JAKE: Don't try and sheet home the blame on me. If you'd been doing your job as a manager you would've kept a far closer eye on her.

DOUGLAS: I didn't want to say it, but it's true, Alex. And we wouldn't be facing a half-million-dollar loss.

ALEX: [*to* JAKE] Someone I like and respect a lot almost died last night. If you can go home and sleep soundly tonight, then you're a full-on sociopath.

JAKE: [*angry, defensive*] Let's get real here. Melissa was reckless, incompetent and stuffed up in a big way. And when you stuff up big-time you get depressed. And depressed people try and kill themselves. It's not my problem, never has been, and don't ever try and lay it on me again.

ALEX: She nearly died.

JAKE: She's a loser. Who fucking cares?!

He storms out. There's a silence.

ALEX: 'Who fucking cares?'. That's the sort of person you want round here?

DOUGLAS: Alex, I'm sorry about this. But could you go to your desk and pack up your personal things and be out of here as quickly as possible?

ALEX *stares at him.*

Sorry. Jake's right. The half-million loss sheets right home to you.

ALEX: [*shocked*] I know this business back to front. You're not going to find someone who can walk in here and do what I can do. [*It suddenly dawns on him.*] Jake? You can't be serious. It'll take years before you get him up to speed.

DOUGLAS: He's the fastest learner I've ever known. Sorry, Alex, but don't hang around talking to people and badmouthing me. Just clear your desk and go.

ALEX: Douglas, you don't need to worry. Whatever's personal that's in my desk the company can have. I'm out of here right now.

He turns and leaves.

Some time later. JAKE *is in Douglas's office again.*

DOUGLAS: I just wanted you to know that Alex's left us.

JAKE: Resigned?

DOUGLAS: I asked him to leave.

JAKE: That blunder *was* his responsibility.

DOUGLAS: Exactly. [*Pause.*] And I also wanted to say that I think you were right to defend yourself against that attack.

JAKE: He wasn't going to get away with it.

DOUGLAS: But the girl nearly did die. I have to say I thought it was a little callous to dismiss her as a loser.

JAKE: Yeah.

DOUGLAS: And say, 'Who cares?'.

JAKE: Yeah. I've felt pretty rotten about it ever since.

DOUGLAS: It was pretty tough.

JAKE: Yeah, it was. I'm glad she pulled through. I've sent her flowers and a card.

DOUGLAS: Jake, I'd like you to step up into Alex's shoes. It's a big ask, but I think you can do it.

JAKE: [*looking suitably surprised*] It is a big ask. I'm really, really honoured that you think I can do it.

DOUGLAS: I wouldn't ask you if I didn't. It's going to be a steep learning curve, but I'll make sure you get every assistance.

JAKE: Douglas. I don't know what to say.

ALEX *sits with* MELISSA, *who is recovering in an easy chair.*

MELISSA: I'm so sorry. It was so stupid.

ALEX: No, I'm sorry. You needed me, and all I could do was get totally and appalling proper.

MELISSA: No. I was the one who did something appalling. Putting you in that position. Like, 'I'm suicidal. Declare your love for me or else.'

ALEX: Can I declare it now, and hope I'm not too late?

There's a pause.

MELISSA: I'd like to wait.

ALEX: Why?

MELISSA: I don't want this to be about pity.

ALEX: It's not pity.

MELISSA: I want to be well. I want to have a new job. I want us to both know what we're doing.

ALEX: It's your call.

MELISSA: Do you understand?

ALEX: No, but I'm going to have to live with it.

MELISSA: Do you understand?

ALEX: No. Because what I feel about you isn't even remotely connected to pity.

MELISSA: Let's just wait.

ALEX: You're looking a hell of a lot better.

MELISSA: I only had one bad moment today. And that was horrible.

ALEX: What?

MELISSA: Jake sent flowers. And a note.

ALEX: There's no way he's got the knowledge, the ability or the energy to do my job. They've saved money by paying him two thirds of what I got, but they'll find it's a false economy.

Six months later. JAKE *enters Douglas's office.*

DOUGLAS: Jake, thanks for coming.

JAKE: Always a pleasure to see you, Douglas.

DOUGLAS: Jake. Your half-yearly review. In terms of financial results, fine.

JAKE: Thank you.

DOUGLAS: Ironically, most of that success can be attributed to SkiFit. We did lose money on that first order, but the way the orders have been rolling in since… [*chuckling*] in a way we've got a lot to thank Melissa for.

JAKE: [*through gritted teeth*] I've worked hard to get those new orders, Douglas.

DOUGLAS: Of course. So anyway, the Board is very pleased with your performance.

JAKE: Good.

DOUGLAS: There do seem to be one or two problem areas.

JAKE: Such as?

DOUGLAS: With product development it's really important to be pro-active. There doesn't seem to be much in the way of innovative new thinking coming out of your department.

JAKE: The two new staff we put on have been pretty disappointing.

DOUGLAS: Their feeling seemed to be that when they came to you for support you don't know quite as much about the detail of our operation as you should.

JAKE: Detail is not my area. I'm the vision guy. Which is what senior management should be. I gave those two guys their brief and I'm deeply disappointed that they're trying to sheet back their inadequacies on me.

DOUGLAS: Jake, the vision thing is vital, but I can't see much evidence you're supplying it, and that's what we're paying you for.

JAKE: You're paying me a fraction of what you paid Alex and expecting me to perform twice as well.

DOUGLAS: More than a fraction, Jake. For someone your age, you're doing remarkably well.

JAKE: There are others who don't think so. Look, I didn't mean to bring this up today, but since it's come up, it's only fair to let you know that I've been made a substantial offer by another firm and I've accepted.

DOUGLAS: [*taken aback*] You've only been in this job for six months and you're out looking for another?

JAKE: I was headhunted, Douglas. They were very impressed by the speed of my promotion.

DOUGLAS: Which was due to me.

JAKE: I have to say that after hearing what I have today, I'm really pleased that I *did* say yes. The salary I'm going to is almost double what I'm being paid here.

DOUGLAS: Jake, I went out on a limb for you.

JAKE: I'll be working for a company that produces something worth producing. Fitness machines? Did you honestly think I was going to spend all my hours boning up on reverse-thread grommet gears or whatever?

DOUGLAS: This is a bloody good business and you should be bloody grateful for what it's given you.

JAKE: It's small beer, Douglas. If you'd had any sense you would've got out years ago. You're never going to make the *Fortune* Top Two Hundred hanging around this dump.

Some days later. DOUGLAS *speaks to* ALEX.

DOUGLAS: I finally saw him in his true colours.

ALEX: Sorry it took you so long.

DOUGLAS: We want you back, Alex. Twenty percent more pay, you name it, but we want you back.

ALEX: Very flattering, Douglas, but I don't think so.

DOUGLAS: Come on, Alex. Don't stand on your dignity.

ALEX: It's about all you left me to stand on, mate.

DOUGLAS: I know you're having a rough time finding a new job out there.

ALEX: Yeah, anyone over fifty is seen as poison.

DOUGLAS: Which is crazy.

ALEX: Not really. You get the youngies cheaper. And work 'em eighty hours a week.

DOUGLAS: Alex, we want you back.

ALEX: I'll think about it.

DOUGLAS: You're living with that girl? Melissa?

ALEX: Yeah. We're very happy.

DOUGLAS: I heard she was working again?

ALEX: Yeah. She got a very good job.

DOUGLAS: I was surprised.

ALEX: Shouldn't have been. She had a name as the gutsy woman who defied her production department and got the hugely successful SkiFit onto the market.

DOUGLAS: I've no doubt in my mind now, that it *was* Jake who was lying.

ALEX: You should thank him. All over the country people are grunting and skiing on the spot.

DOUGLAS: Come back, Alex. Name your terms.

ALEX: I'll think about it. [*To the audience*] I didn't go back. Thought about it, but just couldn't stomach the idea of devoting the rest of my life to dreaming up the next big aerobic thing. Or working under Douglas. As it turned out it was lucky I said no. A month later a big American company got excited about the potential of SkiFit—took over the company, fired Douglas and everyone else, sold off the assets and sent SkiFit offshore to be manufactured in China. I used my severance payout to buy a plant nursery, which makes money and which I enjoy a lot. Seeing things grow is inexplicably satisfying and the daily rhythm of physical labour keeps me fit. Melissa worked out that eighty hours a week wasn't doing much for her or our relationship so we now work together growing things. And we're happy. My boys aren't amused that I'm with someone not much older than them, and my first wife is hugely scornful, but I don't give a hoot. I'm just grateful things have gone as well as they have. When we're not growing things Melissa and I go and climb up real hills, cycle on real bikes, ski across real country and walk along real beaches. The sight of any kind of aerobic device makes us both physically ill. Jake's meteoric rise seems set to continue. Just before he wrecks another firm, the headhunters get him and double his salary. One day he'll do something totally unscrupulous, bankrupt a huge firm, lose squillions for the shareholders, and everyone will say, 'Why the hell didn't someone work out what type he was?'.

THE END

ALSO BY DAVID WILLIAMSON

After the Ball
Amigos
Birthrights/ Soulmates
Brilliant Lies
The Club
Collected Plays Volume I
Collected Plays Volume II
Dead White Males
The Department
Don's Party
Emerald City
Flatfoot
The Great Man / Sanctuary
The Jack Manning Trilogy (Face to Face, A Conversation,
    Charitable Intent)
Money and Friends
The Perfectionist
The Removalists
Siren
Sons of Cain
Third World Blues
Top Silk
Up for Grabs / Corporate Vibes

ABOUT DAVID WILLIAMSON

Brian Kiernan, *David Williamson: A Writer's Career*
David Moore, *David Williamson's Jack Manning Trilogy: A
    Study Guide*

For a full list of our titles, visit our website:

www.currency.com.au

Currency Press
The performing arts publisher
PO Box 2287
Strawberry Hills NSW 2012
Australia
enquiries@currency.com.au
Tel: (02) 9319 5877
Fax: (02) 9319 3649